Scientific
Visualization

K. W. Brodlie
L. A. Carpenter
R. A. Earnshaw

Scientific Visualization

J. R. Gallop
R. J. Hubbold
A. M. Mumford
C. D. Osland
P. Quarendon
(Editors)

Techniques and Applications

With 53 Figures and 15 Tables

Springer-Verlag
Berlin Heidelberg New York
London Paris Tokyo
Hong Kong Barcelona
Budapest

Dr. K. W. Brodlie
University of Leeds
School of Computer Studies
Leeds LS2 9JT, U.K.

L. Carpenter
NAG Ltd.
Wilkinson House
Jordan Hill Road
Oxford OX2 8DR, U.K.

Dr. R. A. Earnshaw
Head of Computer
Graphics
University of Leeds
Leeds LS2 9JT, U.K.

J. R. Gallop
Informatics Division
Rutherford Appleton Lab.
Chilton, DIDCOT
Oxon OX11 0QX, U.K.

Dr. R. J. Hubbold
Dept. of Computer Science
University of Manchester
Manchester M13 9PL, U.K

Dr. A. M. Mumford
Computer Centre
Loughborough University
Loughborough
Leics LE11 3TU, U.K.

C. D. Osland
Central Computing Dept.
Rutherford Appleton Lab.
Chilton, DIDCOT
Oxon OX11 0QX, U.K.

Dr. P. Quarendon
IBM UK Scientific Centre
Athelstan House
St. Clement Street
Winchester
Hants SO23 9DR, U.K.

Front cover picture courtesy of UNIRAS Ltd., and done with UNIRAS agX/VOLUMES

ISBN 3-540-54565-4 Springer-Verlag Berlin Heidelberg New York
ISBN 0-387-54565-4 Springer-Verlag New York Berlin Heidelberg

Library of Congress Cataloging-in-Publication Data
Scientific visualization / edited by K. W. Brodlie ... [etal.] p. cm. "June 23, 1991." Includes bibliographical references and index.
ISBN 3-540-54565-4 (alk. paper). – ISBN 0-387-54565-4 (U.S.: alk. paper)
1. Science – Methodology. 2. Visualization – Technique. I. Brodlie, K. W. Q175.S4242 1992
502.8 – dc20 91-35638

Cover Design: H. Lopka, Ilvesheim
Typesetting: camera-ready by author
33/3140-5 4 3 2 1 0 – Printed on acid-free paper

PREFACE

Background

A group of UK experts on Scientific Visualization and its associated applications gathered at The Cosener's House in Abingdon, Oxfordshire (UK) in February 1991 to consider all aspects of scientific visualization and to produce a number of documents:

- a detailed summary of current knowledge, techniques and applications in the field (this book);
- an Introductory Guide to Visualization that could be widely distributed to the UK academic community as an encouragement to use visualization techniques and tools in their work;
- a Management Report (to the UK Advisory Group On Computer Graphics - AGOCG) documenting the principal results of the workshop and making recommendations as appropriate.

This book proposes a framework through which scientific visualization systems may be understood and their capabilities described. It then provides overviews of the techniques, data facilities and human-computer interface that are required in a scientific visualization system. The ways in which scientific visualization has been applied to a wide range of applications is reviewed and the available products that are scientific visualization systems or contribute to scientific visualization systems are described.

The book is completed by a comprehensive bibliography of literature relevant to scientific visualization and a glossary of terms.

Acknowledgements

This book was predominantly written during the workshop in Abingdon. The participants started from an "input document" produced by Ken Brodlie, Lesley Ann Carpenter, Rae Earnshaw, Julian Gallop (with Janet Haswell), Chris Osland and Peter Quarendon. Together with Anne Mumford, this team formed the Organizing Committee for the workshop.

The document was refined by subgroups during the workshop and the resulting status report was transformed into this book by the editors.

This workshop could not have taken place without the considerable enthusiasm of the participants. Their willingness to exchange views and work hard over the four days was very much appreciated. The correspondence between people and sections of the final book is as follows:

Malcolm Austen Framework, Bibliography, several illustrations (*Oxford University Computing Service*)

Neil Bowers Framework, Human-Computer Interface, information on X11 toolkits, data compression, data formats (*Computer Science Department, University of Leeds*)

Ken Brodlie Editor of Techniques Chapter, Products (*Computer Science Department, University of Leeds*)

Alex Butler Data Facilities, Human-Computer Interface (*Centre for Novel Computing, University of Manchester*)

Lesley Carpenter Editor of Data Facilities Chapter, Human-Computer Interface, contribution to Framework (*NAG Ltd, Oxford*)

Syd Chapman Framework, sections in Applications (*IBM UK Scientific Centre, Winchester*)

Kate Crennell Data Facilities, Applications, section on Image Processing (*Neutron Instrumentation Division, Rutherford Appleton Laboratory*)

Rae Earnshaw Editor of Introduction and Conclusions chapters (*Computing Service, University of Leeds*)

Todd Elvins Techniques, section on 3D scene description, several illustrations (*San Diego Supercomputer Center*)

Terrance Fernando Data Facilities, Applications, sections on database products (*Computer Science Department, University of Leeds*)

Mark Fuller Framework, Human-Computer Interface, initial Bibliography (*VisLab, Department of Computer Science, University of Sheffield*)

Julian Gallop Book Editor, Editor of Products chapter and Enabling Technologies Appendix, Data Facilities (*Informatics Department, Rutherford Appleton Laboratory*)

Phil Gardner Techniques (*Computing Services, Clinical Research Centre*)

Janet Haswell Techniques, Products, Bibliography, initial Products chapter and sections of Enabling Technologies (*Informatics Department, Rutherford Appleton Laboratory*)

Hilary Hearnshaw Framework, Human-Computer Interface (*Midlands Regional Research Laboratory, University of Leicester*)

Terry Hewitt Techniques, Products (*Computer Graphics Unit, University of Manchester*)

Bob Hopgood Human-Computer Interface, contribution on Framework (*Informatics Department, Rutherford Appleton Laboratory*)

Roger Hubbold Editor of Human-Computer Interface chapter (*Computer Science Department, University of Manchester*)

David Knight Techniques, Applications, several illustrations (*Computing Laboratory, Oxford University*)

Chris Little Data Facilities, sections on Applications, several illustrations (*Central Computing, Meteorological Office*)

Roy Middleton Framework, Products (*Edinburgh University Computing Service*)

Anne Mumford Host to the Workshop, Data Facilities (*Computer Centre, Loughborough University*)

Adel Nasser Data Facilities, Applications, section on Enabling Technologies (*Computer Graphics Unit, University of Manchester*)

Chris Osland Book Editor, Editor of Framework Chapter, Glossary and Bibliography (*Central Computing Department, Rutherford Appleton Laboratory*)

Rajka Popovic Framework, Bibliography (*Informatics Department, Rutherford Appleton Laboratory*)

Peter Quarendon Editor of Applications chapter, Techniques (*IBM UK Scientific Centre, Winchester*)

Mahes Visvalingam Framework, Human-Computer Interface, section on Applications (*Department of Computer Science, University of Hull*)

Howard Watkins Techniques, Products, section on Applications, several illustrations (*Exploration Consultants Limited (ECL), Henley-on-Thames*)

David Watson Data Facilities, sections on Applications, contribution to Framework (*IBM UK Scientific Centre, Winchester*)

Norman Wiseman Data Facilities, Products (*NERC Computer Services, Keyworth*)

Michael Wood Techniques, Human-Computer Interface, section on Applications (*Department of Geography, University of Aberdeen*)

We would like to thank Roger Hubbold for taking on the chairing of the Human-Computer Interface group which started the workshop with a clean sheet of paper; also Norman Wiseman who was originally part of the organizing team.

The rapid production of this book has been made possible by the continuing efforts of those at the workshop and the many colleagues of the editors who have proof-read parts of the book, especially Adrian Clark of the University of Essex.

The staff at Cosener's House in Abingdon ensured the smooth running of the domestic arrangements which were excellent. Sheila Davidson of the Rutherford Appleton Laboratory and Joanne Barradell of Loughborough University assisted in the administration of the event; Janet Haswell, Rajka Popovic and Chris Seelig in the technical arrangements; all of these people deserve our considerable thanks. In addition, Chris Seelig kindly prepared the figures illustrating the Framework model and Nigel Diaper of Rutherford's Reprographics Department provided invaluable assistance, in and out of normal hours, with the final typesetting of the book.

For the duration of the workshop, a number of suppliers were demonstrating various visualization products. The organizers of the workshop would like to thank the following for the equipment and software they lent, which acted as a catalyst for discussion.

DEC UK
Dynamic Graphics
Fairfield Imaging
IBM (UK) Scientific Centre
Silicon Graphics
Stardent Computer UK
Sun Microsystems (UK)
Wavefront Technologies UK

We especially thank Chris Osland and Julian Gallop for all their work in the typesetting of this book and the production of the final manuscript. This has facilitated the rapid execution of a series of major edits and the incorporation of additional material subsequent to the workshop.

K. W. Brodlie
L. A. Carpenter
R. A. Earnshaw
J. R. Gallop
R. J. Hubbold
A. M. Mumford
C. D. Osland
P. Quarendon

Abingdon, June 1991

COPYRIGHT MATERIAL

We are indebted to the following for the use of copyright material:

IBM (UK) Scientific Centre, Winchester for permission to reproduce the visualizations of the pottery distributions at Fairoak (Figure 38 on page 144);

San Diego Supercomputer Center for permission to reproduce the figure illustrating isosurfaces (Figure 23 on page 59);

UK Government for permission to reproduce the Crown Copyright figures for the cartography section (Figure 35 on page 138) and the meteorology section (Figure 45 on page 158 and Figure 46 on page 159).

DISCLAIMER

Views expressed in this book reflect those of one or more of the participants at the workshop, not necessarily a majority of those involved. Value judgements are offered in good faith from experience of products, systems and services by the participants. The editors and publishers do not hold themselves responsible for the views expressed by the participants through this book.

TRADEMARKS

In this book, it was essential to use terms that may be trademarks or registered trademarks in order that products might be identified unambiguously. The following are known to be used (tm) or registered (rtm) by the following companies; we trust that others that may not be noted are known to readers and are referenced in a manner acceptable to the companies concerned.

Adobe Inc
PostScript (rtm)

American Telephone and Telegraph Company
UNIX (rtm), Pixel Planes

Apple Computer Corporation
Macintosh

Bradley Associates
GINO-F (tm)

Compuserve Inc
GIF (tm)

International Business Machines
IBM, PC, PS/2, GDDM (tm)

Leading Technology Products
TekBase

Massachusetts Institute of Technology
X Window System (tm), X11

Numerical Algorithms Group Ltd and Numerical Algorithms Group, Inc. (NAG)
NAG (rtm)

Ohio Supercomputer Center
apE

Oracle Corporation
ORACLE

Pixar
RenderMan (tm)

Precision Visuals Inc
PV-WAVE

Relational Technology Inc.
INGRES

Stardent Computer
AVS (tm), Doré

Sun Microsystems
OpenWindows (tm), SparcStation (tm), SunView (tm), SunVision (tm), XDR (tm), XGL (tm)

Tektronix Inc
PLOT-10 (rtm)

Vital Images Inc
VoxelView (tm)

VPL
DataGlove

Wavefront Technologies Inc
Advanced Visualizer (tm), Data Visualizer (tm)

ABOUT THE EDITORS

K. W. Brodlie

Ken Brodlie is Senior Lecturer in the School of Computer Studies at the University of Leeds, and Deputy Head of the Division of Computer Science. He has had a long involvement with international standards for computer graphics, and presently chairs the group looking at the revision of GKS. A special interest has been the validation of implementations of graphics standards, where he has contributed to the European-wide testing services for GKS and CGI implementations. His interest in scientific visualization began with an involvement in the NAG Graphics Library, and has extended to research into new visualization systems that integrate numerical computation and graphics. He was the founding chairman of the UK Chapter of Eurographics, and is on the Editorial Board of Computer Graphics Forum.

L. A. Carpenter

Lesley Ann Carpenter is a full time employee of the Numerical Algorithms Group (NAG Ltd), Oxford, a non-profit making Company specializing in the provision of highly portable numerical and statistical software, for which it has a world-wide reputation. At NAG, she heads the Visualization Group and is also Deputy Divisional Manager of the Software Environments Division. She has strong interests in computer graphics, software integration and scientific visualization of numerical computation. She is a long-standing member of the Eurographics Computer Graphics Association and actively involved in their Working Group on Scientific Visualization. She is joint Technical Coordinator of the GRASPARC project, a collaborative visualization project between NAG Ltd, Leeds University and Quintek Ltd. She is also a member of the Computer Society of the IEEE, Technical Committee on Computer Graphics (TCCG) and is involved in work on Computer Graphics standardization in the United Kingdom (BSI). She has substantial experience in the field of Computer Graphics (over

12 years) having worked initially in the field of oil exploration and subsequently being responsible for the development and support of software for a very wide range of graphical software packages, protocols and hardware.

R. A. Earnshaw

Rae Earnshaw is Head of Computer Graphics at the University of Leeds with interests in graphics algorithms, integrated graphics and text, and scientific visualization. He has been a Visiting Professor at Illinois Institute of Technology, George Washington University, and Northwestern Polytechnical University, China. He has acted as a consultant to US companies and the College CAD/CAM consortium and given seminars at UK and US institutions and research laboratories. He is a Fellow of the British Computer Society (BCS) and Chairman of the Computer Graphics and Displays Group. He was a Director of the 1985 Advanced Study Institute on "Fundamental Algorithms for Computer Graphics", and Co-Chair of the BCS/ACM International Summer Institute on "State of the Art in Computer Graphics" held in Scotland in 1986.

J. R. Gallop

Julian Gallop is head of the Visualization Group in the Informatics Department at Rutherford Appleton Laboratory. He participated in the ISO standardization of computer graphics. His early interest there was GKS; he was one of the document editors of the standard and co-authored an introductory text that sells in the UK and USA and has been translated into other languages. More recently he has participated in the standardization of computer graphics language bindings and PHIGS PLUS. His present interests are the visualization of scientific and engineering data, image processing and, via a CEC (Commission of the European Community) Esprit II project, the integration of graphics and communications.

R. J. Hubbold

Roger Hubbold is a Senior Lecturer in Computer Science at the University of Manchester, and Associate Director, responsible for Visualization, in the Centre for Novel Computing. He obtained his B.Sc. and Ph.D. degrees from the University of Leicester in 1967 and 1971 respectively. He has been involved in computer graphics since 1967, in universities and in industry, and from 1974 until 1987 was the first director of the University of Manchester Computer Graphics Unit. In 1985 he moved to his present post to concentrate on longer-term research in graphics and visualization. His current research is on software architectures for interactive visualization using parallel computing systems. He has published and lectured in the USA, Japan, Australia and throughout Europe on interactive computer graphics and visualization, and is co-author of a leading text book on the PHIGS and PHIGS PLUS graphics standards. He is a Fellow of the European Association for Computer Graphics, holding positions as Chair of its Conference Board, Vice-Chair, and then Chair of the Association in 1989/90.

A. M. Mumford

Anne Mumford is the Coordinator of the UK's Advisory Group on Computer Graphics. This has been set up to advise the UK academic funding bodies on strategies and purchases for computer graphics support as well as raising awareness of new issues and being concerned with training initiatives. Dr Mumford is based at Loughborough University where she had responsibility for computer graphics support before being seconded in 1989 to work on the newly set up AGOCG. Dr Mumford has had a long standing interest in computer graphics standards. She has been involved as a UK delegate to international standards meetings since 1985 and is the chair of the ISO Computer Graphics Metafile (CGM) standard working group. She organized the UK CGM demonstration which ran at Eurographics UK and the Computer Graphics Show in 1989. Dr Mumford has written widely about the subject of graphical data storage in general and CGM in particular. She has co-authored (with Lofton Henderson) the Butterworth-Heinemann book on The Computer Graphics Metafile and is European Series Editor for their graphics standards series. Dr Mumford is Chairman of Eurographics UK.

C. D. Osland

Chris Osland is Head of Computer Graphics at the Rutherford Appleton Laboratory (part of the UK Science and Engineering Research Council). His principal interests are the integration of computer graphics with video and the improvement of all forms of presentation by learning from the graphics arts. He was in charge of the ISO group that developed the Computer Graphics Metafile ISO standard from 1981 to 1985 and has lectured extensively on this and other graphics topics. Since 1988 he has designed and implemented the Rutherford Video Facility, providing direct video output for UK scientists and engineers.

P. Quarendon

Peter Quarendon has been involved in graphics at IBM for 15 years, originally prototyping and then producing a product, GDDM, to make graphics available on all IBM mainframe terminals. He moved to the IBM UK Scientific Centre in 1983, to carry out research in the scientific uses of computer graphics. He developed WINSOM, intended as an easy-to-use package for scientific rendering and has been involved in numerous visualization projects. It jointly won the BCS computer applications award in 1989. After a period as manager of the Graphics Applications Research Group, he is now Graphics Research Leader at the Scientific Centre and responsible for technical aspects of the graphics and visualization research.

CONTENTS

APPENDICES

ILLUSTRATIONS

TABLES

Chapter 1

INTRODUCTION

Edited by Rae Earnshaw

1.1 What is Scientific Visualization?

Scientific Visualization is concerned with exploring data and information in such a way as to gain understanding and insight into the data.

"The purpose of computing is insight, not numbers", wrote the much cited Richard Hamming [Hamming62]. The goal of scientific visualization is to promote a deeper level of understanding of the data under investigation and to foster new insight into the underlying processes, relying on the human's powerful ability to visualize.

To achieve this goal, scientific visualization utilises aspects in the areas of computer graphics, user-interface methodology, image processing, system design, and signal processing. Formerly these were considered independent fields, but convergence is being brought about by the use of analogous techniques in the different areas. The term Visualization in Scientific Computing (or ViSC) [McCormick87] is also used for this field, or quite simply visualization.

Visualization highlights applications and application areas because it is concerned to provide leverage in these areas to enable the user to achieve more with the computing tools now available. In a number of instances, the tools and techniques of visualization have been used to analyse and display large volumes of, often time-varying, multidimensional data in such a way as to allow the user to extract significant features and results quickly and easily.

Such tools benefit from the availability of modern workstations with good performance, large amounts of memory and disk space, and with powerful graphics facilities - both in terms of range of colours available and also speed of display. This close coupling of graphics and raw computational performance is a powerful combination for those areas where visual insight is an important part of the problem-solving capability.

Workstations can now offer substantial computational power coupled with high speed 3D graphics pipeline, which can transform the processed data into graphical images, often in real time. These facilities can be exploited to significant advantage in application areas such as modelling, simulation, and animation. Real-time dynamical simulation can involve the processing and display of large amounts of data, and often the only effective analysis of the performance or validity of the model is through visual observation.

In those cases where additional computational resource is required, the calculation can be sent to a supercomputer (or other advanced workstations with spare capacity) and the resulting image retrieved for viewing, and perhaps interaction, when ready.

These advances will encourage the study of increasingly complex and detailed mathematical models and simulations. This results in a closer approximation to reality, thus enhancing the possibility of acquiring new knowledge and understanding. Scientific visualization is concerned with methods of generating and presenting large collections of numerical values containing a great deal of information. The scientist has to be able to make effective use of this information for analytic purposes.

A further aspect is that increases in computer performance allow 3D problems in simulation and design to be done interactively. In addition, processes that formerly separated out simulation and design can now bring them together (e.g. in Computer Aided Design (CAD), design of new drugs etc). This in turn moves the user into a new era of design methodology.

Fine control over simulations and interactivity, and use of increased computer performance mean that vast amounts of multi-dimensional data can be generated. Superworkstations allow this data to be displayed in optimum ways. These features and capabilities are driving the current wave of interest in scientific visualization.

A further current trend is to make software tools for visualization more user-friendly and accessible to a wide variety of application areas, thus increasing their potential and usability.

Nielson et al [Nielson90] contains a wide variety of current applications of scientific visualization and also an excellent bibliography of scientific papers. A video tape is can also be obtained to complement the material presented in this book.

Frenkel [Frenkel88] provides a general introduction to basic visualization techniques.

Thalmann [Thalmann90] contains a number of papers in the areas of scientific visualization and graphical simulation.

1.2 History and Background

A report published in 1987 [McCormick87] recommended a series of short term and long-term initiatives. A further Report [DeFanti89] by the same authors detailed the progress made between 1987 and 1989. The authors (Bruce McCormick, Tom DeFanti, and Maxine Brown) reported the outcome of a two-day workshop on "Visualization in Scientific Computing" held in February 1987. This was held as a result of an earlier Panel Meeting on "Graphics, Image Processing and Workstations" sponsored by the Division of Advanced Scientific Computing of the National Science Foundation (NSF). The workshop was attended by representatives of Academia, Industry, and Government Research Laboratories.

McCormick et al [McCormick87] summarises the conclusions and recommendations. Here are a number of the principal conclusions.

1. High Volume Data Sources

   Data sources such as supercomputers, satellites, spacecraft, radio astronomy arrays, instrument arrays and medical scanners all produce large volumes of data to be analysed. The numbers and density of such data sources are expected to increase as technology moves forward. Satellites have resolutions 10-100 times higher than a few years ago. Terabyte data sets are becoming increasingly common in systems concerned with real-time recording of data. Scientific visualization is an essential tool when handling such large amounts of data.

2. The Value of Interdisciplinary Teams

   Systems concerned with scientific visualization benefited from having collaborating disciplines. For example, computational scientists and engineers could combine in areas such as fluid dynamics and molecular modelling; visualization scientists and engineers could combine in areas concerned with visualization software, hardware, and networking; artists and cognitive scientists could ensure that the best forms of visual communication were used - colour, composition, visual representation, visual perception, etc..

3. Visualization Issues for Tool Makers

   The following areas were identified as needing development and support for the future -

   - Interactive steering of computations and simulations
   - Workstation-driven use of supercomputers
   - Graphics-oriented programming environments
   - Visualization of higher-dimensional scalar, vector, and tensor fields
   - Dynamic visualization of fields and flows
   - High bandwidth networks for pictures
   - Handling terabyte data sets - for signal and image processing
   - Vectorized and parallelized algorithms
   - Specialized architectures for graphics and image processing
   - Establishing a framework for international standards in scientific visualization

4. Benefits of Scientific Visualization

   The following potential benefits were identified -

   - Integrated set of portable tools
   - Increased scientific progress and collaboration
   - Increased scientific productivity
   - Standardization of tools
   - Improved market competitiveness
   - Improved overall usefulness of advanced computing facilities

The following were some of the principal recommendations of the Report -

1. Develop new and useful tools for the future
2. Distribute tools to provide opportunities for use
3. Greater funding support needed (e.g. as a percentage of national expenditure)
4. Fund both research (tool users) and technology (tool makers)
5. Fund immediate and long term provision for scientific visualization
6. Enhance scientific and engineering opportunities
7. Recognise the short-term potential of visualization environments
8. Address the long term goals of visualization environments
9. Address the issue of industrial competitiveness

In addition, a range of application areas in science, engineering, and medicine were presented as illustrative of the current uses of scientific visualization tools.

1.3 Current Activities in Scientific Visualization

1.3.1 USA

The principal recommendations of the McCormick Report were that national funding should be provided for short and long term provision of tools and environments to support scientific visualization, and to make these available to the scientific and engineering community at large. Such provision was considered to be essential if the enabling tools were to be effectively harnessed by current and future scientists and engineers.

Such tools often require access to significant computational resources. A natural focal point for these developments has been the funding of National Supercomputer Centres - to provide both the facilities and access to them by the community.

An example of this at the San Diego Supercomputer Center is the development of network-based general purpose visualization tools. These are accessed by 2800 users with around 350 different applications. Such users access the facility by a variety of different routes including dial-in lines, national networks, and dedicated high-speed links. In addition to this broad range of provision there are also more specialized tools for high-end applications (e.g. molecular modelling, computational fluid dynamics).

Similar provision has also been made at other Supercomputer Centers at Cornell, Pittsburgh, and the University of Illinois at Urbana-Champaign.

Workshops on Scientific Visualization have been established by ACM SIGGRAPH and IEEE to address specific aspects such as Data Facilities (to facilitate ease of use and transfer of information), and Volume Visualization (to enable representation of real 3D information and to give interior views). Representatives from the Department of Defense and the Department of Energy have initiated a Working Group to define a Visualization Reference Model.

In addition, a large number of major Universities are establishing Visualization Laboratories, and often such installations receive supplementary funding for further proposals in specific application areas. Funding is provided by such bodies as NSF, DARPA, and NASA. State supercomputers and associated visualization facilities exist in Ohio, North Carolina, Minnesota, Utah, Alaska, and Florida.

To provide a forum for the presentation and discussion of the latest advances in Scientific Visualization, the IEEE Technical Committee on Computer Graphics has established an international visualization conference, which is held on an annual basis.

In addition, the NSF is providing funds for the support and promotion of educational initiatives in Scientific Visualization by means of Institutes, Workshops, and Summer Schools.

Fast networks are required for distributed and remote visualization. Developments in networking infrastructure are planned to provide faster communication, interconnection, and the ability to aggregate computing resources at different locations on to one particular problem. For example, the CASA test bed project is funded by the NSF to develop a 1 gigabit/second network link between Los Alamos National Laboratory, California Institute of Technology, and San Diego Supercomputer Center, to enable all three resources to be concentrated on one application simultaneously.

The problems of graphics and networking were discussed at a joint SIGGRAPH/SIGCOMM workshop in January 1991. The issues were discussed in the light of the demands of visualization applications. Further joint discussions are planned.

A recent multi-million dollar grant has recently been awarded by NSF to California Institute of Technology, Brown University, University of Utah, Cornell University, and the University of North Carolina at Chapel Hill, to explore the foundations of computer graphics and visualization.

1.3.2 UK

A number of centres in UK academic institutions are concerned with application areas such as molecular modelling and computational fluid dynamics (CFD). There are a number of collaborative projects between academia and industry in the areas of parallel processing and scientific visualization. One example, GRASPARC, a GRAphical environment for Supporting PARallel Computing, is a joint project between the Numerical Algorithms Group Ltd. (NAG), the University of Leeds (School of Computer Studies), and Quintek Ltd. The major objective of the work is to improve the interaction between the numerical analyst and the parallel computer through the development of interactive visualization software.

The IBM UK Scientific Centre in Winchester is primarily concerned with scientific visualization and has a Visualization Group, a European Visualization Group, a Medical Imaging Group, and a Parallel Programming and Visualization Group. There are a number of collaborative projects with academia and industry in the areas of parallel processing, user-interface aspects, and medical informatics.

NERC has a Visualization Advisory Group concerned with evaluating products for the areas of geological surveys and oceanography. SERC, as early as 1979, set up the Starlink Project for the coordination of astronomical computing and this has generated a large base of data analysis software, enabling astronomical data to be visualized.

Informatics Department at the Rutherford Appleton Laboratory (RAL) is responsible for visualization on behalf of the SERC Engineering Board's computing community, providing facilities and assessments - including an evaluation of superworkstations in the areas of hardware and software. RAL Central Computing Department has developed a video facility for use by the academic and research community in the UK, and this has been used for visualization in the areas of oceanography, atmospheric physics, mechanical engineering, ecological simulation, and computational fluid dynamics.

There are many groups which are university based, of which an example is the Sheffield VisLab, which is a University-wide visualization facility.

The AGOCG Scientific Visualization Workshop which produced this book arose out of an initiative by the Computer Board and AGOCG.

In general, there is insufficient funding for this field and most of the groups are small.

1.3.3 Europe

IBM has a number of European centres actively involved in projects involving Scientific Visualization. These include the European Petroleum Applications Centre (EPAC) in Bergen, the Paris Scientific Centre which is involved in visualization in the medical area, and the

European Scientific Centre in Rome which is involved in engineering and modelling turbulent flow. IBM also has a joint project with the Centre of Competence in Visualization at the University of Aix-Marseilles.

In early 1986 the Physics Analysis Workstation (PAW) Project was started at CERN, aimed at exploiting personal workstations.

FhG-AGD in Darmstadt is working on a number of areas, including tools for volume visualization on a variety of platforms, and handling different kinds of data sets.

Delft University of Technology has research interests in the area of scientific visualization.

Eurographics arranged a Workshop on Scientific Visualization in April 1990 at Clamart (Paris, France). The proceedings will be available from Springer-Verlag. A further workshop was held in April 1991 at Delft (the Netherlands).

However there is no direct funding for visualization in European Community programmes.

1.4 Background to the AGOCG Workshop

The Advisory Group on Computer Graphics (AGOCG) has been set up to play a number of roles in the academic and research community in the UK. Its terms of reference are:

- to advise the Computer Board and the Research Councils on all aspects of computer graphics
- to be aware of advances in computer graphics in both the standards area and in innovative new technology in both hardware and software
- to liaise with the community to help identify requirements in the area of computer graphics
- to recommend to all relevant funding bodies options for purchase, support and development that would improve the environment available to the academic community in the area of computer graphics

- to ensure that facilities for education and training are provided to the community on the benefits arising from the use of computer graphics including standards.

It was with all these roles in mind that it seemed the right time to gather together experts to look at the approach which should be taken by AGOCG in the area of scientific visualization.

AGOCG organized a workshop (22-25 February 1991) which was supported by the UK Computer Board and Eurographics UK. This brought together 29 experts from UK academic and research establishments and from industry and Todd Elvins, a visualization programmer, from the San Diego Supercomputer Center. The participants came from a range of disciplines and from research and support environments. This led to a very useful exchange of ideas.

The workshop produced a number of outputs:

- a **Management Report**, informing AGOCG of the status of results of the workshop and making recommendations concerning its support and development;
- an **Introductory Guide to Visualization**, which is to be made widely available in the UK academic and research community by AGOCG;
- a **Status Report on Scientific Visualization** (this book);
- a **Video Tape**, containing examples of visualization techniques and commentary explaining their use.

The purpose of the separate Introductory Guide is to explain in simple terms what scientific visualization is and what it can do, and give illustrations and explanations of the technical terms in a way the non-specialist can understand. It is intended for the general reader in a department or Computer Centre, as well as the scientific specialist, and provides a general outline of the benefits and possibilities of scientific visualization.

1.5 Introduction to the Book

This book is the main output of the workshop. It is a full consideration of the subject of scientific visualization and will serve as both a contribution to the field and as a reference guide within the community. The main subjects covered include:

- **Framework**
- **Visualization Techniques**
- **Data Facilities**
- **Human-Computer Interface**
- **Applications**
- **Products**

A general introduction and conclusion plus supplementary material on **Enabling Technologies**, a **Glossary** and an extensive **Bibliography** are also included. This is valuable for the professional in the field as a reference guide. It will also provide an evaluation of applications and products for future purchase recommendation and as the basis for further evaluation.

This book was drafted by the participants at the workshop based on input documents produced by the organizing committee prior to the meeting for all sections except the human-computer interface section.

1.6 Recommendations of the Workshop

The major recommendations of the workshop are twofold:

- Scientific visualization systems should be made widely available to a range of disciplines in the UK academic and research community.

- The concepts and benefits of scientific visualization should be made widely understood in the UK academic and research community.

These two recommendations lead to a set of specific recommendations concerning product evaluation, purchase and education and training to enable these to be implemented. These are:

1. Evaluations Leading to Proposals to AGOCG

   Further evaluations and investigations are required to advise AGOCG on the solutions available for the provision to the community of scientific visualization in the short, medium and long term. The workshop set the framework for such evaluations and specific work needs to be addressed by AGOCG. This includes:

   a. Further work should be undertaken on the viability, merits, operational environments of the various distributed strategies for visualization systems. The results of this work should be distributed widely via an AGOCG Technical Report. The study should investigate the different hardware/software scenarios: this would include the use of dataflow systems (such as AVS, Khoros and apE) in distributed computing environments for different applications and the use and accessibility of supercomputer facilities, particularly from remote sites.

b. Recognising that recently funded parallel processing initiatives have been set up in the UK, work should be undertaken to make the current visualization systems available on these machines. Further research into graphics and visualization systems for parallel applications is needed.

c. Evaluation of the various formats for generic application data and image transfer needs to be made and recommendations made.

d. Following the evaluations described in these recommendations AGOCG should extend the current UK Computer Board/CCTA Graphics Operational Requirement to include the needs of scientific visualization which are addressed by the other recommendations of this workshop.

e. There is a need to investigate how scientists solve their problems and how scientific visualization can assist.

f. Studies should be undertaken to consider alternative methods for interaction - for example how to process multi-modal, multi-device dialogues - in relation to scientific visualization.

2. Support, Education and Awareness

a. Support needs to be given to this area and it is recommended that both the Computer Board and the SERC allocate one man year per year for 3 years to support general queries about systems and developments in this area. Other research councils and funding bodies are encouraged to allocate resources to this effort. Close working between these support people is essential.

b. It will be necessary to provide education in the use of these systems. AGOCG should recommend Computer Board funding within their Training Initiative for training, information exchange and development of training materials based on purchases and recommendations made over the period of the post in this area.

3. General

   a. There were both academic and industrial representatives at the workshop. There is a clear need to foster these links and to continue the exchange of insights and knowledge in this area in future collaborative work which should be encouraged and fostered by all parties.

   b. In the light of the considerable implications of scientific visualization for networking, it is recommended that AGOCG forward the workshop report to the Joint Network Team (JNT) for their comments.

1.7 Key References

Most references are provided in the bibliography at the end of the book. A few references which are considered key for this chapter are provided here.

[Frenkel88]

Frenkel K. A., "The Art and Science of Visualizing Data", *Communications of the ACM* vol 31 (2), **(1988)**, *pp 110-121.*

An introductory paper which looks at a range of application areas and the uses of visualization tools and techniques. A wide range of pictures illustrate some of the techniques currently being used.

[Hamming62]

Hamming R.W., "Numerical Methods for Scientists and Engineers", **(McGraw-Hill, New York, 1962)**.

Introduction to numerical techniques for those concerned with application areas.

[McCormick87]

McCormick B., DeFanti T.A., Brown M.D., "Visualization in Scientific Computing", *ACM SIGGRAPH Computer Graphics* vol 21 (6), **(Nov 1987)**.

The original Panel Report which outlines the political, economic, educational, and technological aspects of scientific visualization as an emerging discipline.

[DeFanti89]

DeFanti T.A., Brown M.D., McCormick B.H., "Visualization - Expanding Scientific and Engineering Research Opportunities", *IEEE Computer* vol 23 (8), **(Aug 1989)**, *pp 12-25.*

An updated version of the original McCormick report [McCormick87], outlining current progress and advances in scientific visualization.

[Nielson90]

"Visualization in Scientific Computing", ed. Nielson G.M., Shriver B., Rosenblum L.J., ISBN 0-8186-8979-X, **(IEEE Computer Society Press, 1990)**.

A collection of papers in the areas of techniques and applications of scientific visualization from a variety of academic, government, and industrial organizations in the USA.

[Thalmann90]

"Scientific Visualization and Graphics Simulation", ed. Thalmann D., ISBN 0-471-92742-2, **(John Wiley, 1990)**.

Computational and graphical techniques that are necessary to visualize scientific experiments are surveyed in this volume, with a number of case studies in particular application areas.

Chapter 2

FRAMEWORK

Edited by Chris Osland

2.1 Introduction

There are many different meanings of the word ***visualization***. Outside the domain of computer graphics, accepted definitions include

- making visible, especially to one's mind (thing not visible to the eye) [OED69],
- forming a mental image of something (thing not present to the sight, an abstraction, etc.) [Webster70].

The term ***Scientific Visualization*** (or ***Visualization for Scientific Computing***, commonly abbreviated to ***ViSC***) is defined in more specific terms:

- the use of computer imaging technology as a tool for comprehending data obtained by simulation or physical measurement [Haber90];
- [the] techniques that allow scientists and engineers to extract knowledge from the results of simulations and computations [Nielson90].

Clearly the first two definitions refer to the human activity of visualization whereas the second two refer to the use of computing technology to assist this activity.

The differences between the definitions in each group are those of stress than of substance, and are due to the authors choosing wider or narrower scopes for their purposes. In order that discussion of visualization at the AGOCG workshop could be well-focussed, it was therefore essential for a particular scope to be decided.

Once this had been done, a framework that described visualization systems in abstract terms was evolved and the techniques, data facilities, products and applications were analysed in these terms.

This framework is described in terms of a number of models, from long-established models of the process of creative thinking, through a formal model of the various components of a visualization system, down to the fine detail within some of these components.

2.2 Scope

It has been common [Haber90] to regard visualization conceptually as a ***complex of generalized mappings of data from raw data to rendered image***. Indeed, there are suppliers of what could be regarded as purely graphical systems that use the term ***visualization*** synonymously with ***graphics*** for the purposes of publicity.

If the term ***visualization*** is to refer to processes that provide "insight into numbers" [Hamming62] it must include not just the activity of producing pictures from data, but also the complete cycle of processes that enables the data to accessed and manipulated.

2.3 High Level Models

2.3.1 *Model of Creative Thinking*

Scientific Visualization systems form part of the process of scientific enquiry. Wallas [Wallas26] identified four stages in creative thinking.

- Preparation
- Incubation
- Illumination
- Verification

The processes of incubation and illumination are mental processes which form an intrinsic part of creative visualization. Scientific Visualization systems facilitate the processes of preparation and verification. A similar model to Wallas' was developed by Helmholz and Poincaré in the early 1900s and refined by Getzels in the 1960s [Getzels80].

Visualization of data in a variety of ways can form part of a brainstorming exercise which reveals anomalies or unexpected trends and patterns during the preparation stage. Verification is also facilitated by the ease with which a hypothesis, based on some initial insight, may be explored through visualization of data.

2.3.2 *Model of Scientific Investigation*

Watson has proposed [Watson90] a cyclic model of the process of scientific investigation, shown in Figure 1 on page 17.

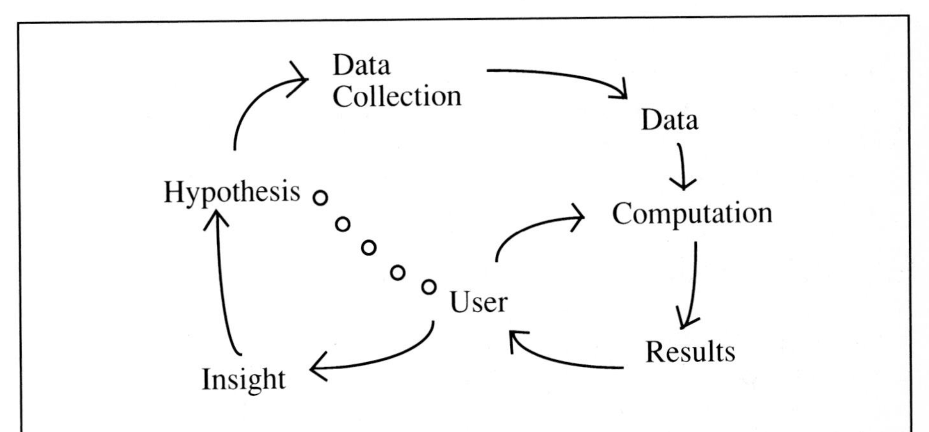

Figure 1. Watson's model of scientific investigation: This model includes both the human activities and the computer-assisted activities involved in Scientific Visualization

He postulates a cyclic process with the scientist using insight to construct a hypothesis, collect data, control a computation, view and interpret the output and then repeat and refine the cycle. While not the only possible model, this is an appropriate model for the external activity implied in the Preparation and Verification processes of Wallas' model. The computational and analysis aspects are further explored in [Upson89b] and [Carpenter91].

We believe it is useful to include all aspects of Watson's model when discussing visualization systems. For this reason the framework model described below includes components (modules) corresponding to all aspects of the Watson model, except that "Insight", "Hypothesis" and "User" are combined into one module ("The User").

The functional aspects that are specific to visualization are discussed in detail in Chapter 3 (Techniques) and Chapter 4 (Data Facilities). Other functional aspects common to many other computational disciplines are referred to as "Enabling Technologies" and are described in more detail in Appendix A.

In addition to the functional capabilities of a visualization system which can be analysed in terms of this framework model, there are many qualitative aspects that are crucially important to the suitability of a visualization system for its application. The Human-Computer Interface (HCI) is noted briefly in this chapter and analysed in more detail in Chapter 5. Other qualitative aspects include responsiveness, accuracy of representation, process time versus cost, appropriateness of tool sets, help and education.

2.3.3 *Detailed Model of Visualization*

At the 1990 IEEE conference on Visualization, Haber and McNabb [Haber90] provide a detailed model of part of the visualization process. They distinguish three subprocesses:

- Data Enrichment / Enhancement
- Visualization Mapping
- Rendering

These successively act upon "simulation data" until the result of the rendering is a "displayable image". The term ***abstract visualization object*** is used to describe an imaginary object in time and space that is the result of the visualization mapping, before rendering it into an image. This provides fine detail of the visualization process in the area closest to rendering.

In this book, the term "Internal Data" is used rather than "Simulation Data", since the latter appears to exclude physical measurements, which are as likely to be the subject of a visualization process as simulation data.

Similar models were used in the development of apE [Dyer90]. The developers of apE stress the need for steering and introduce a limited form of feedback between interpretation and simulation phases.

The philosophy underlying these models is rather restrictive. It seems to suggest a one-way progression from internal data to rendered image, ignoring the cyclic process proposed by Watson.

More generalized network models for the structuring of the transformation models are used in Khoros [Rasure91] and AVS [Upson89a]. The design aim of AVS was for users to be able to interact with their problems. It was one of the earliest attempts to reach beyond the traditional view of visualization comprising pictorial output. In AVS, users can interact with data input and data transformation as well as rendering.

The underlying visualization model is based on the analysis of a number of mathematical problems and the mechanisms which are required to effect a complete solution. The concept of a computational cycle is introduced: although a logical sequence of events is implied, the opportunity exists for the user to back-track to a previous step (or steps), to modify perhaps an input parameter or data value and to repeat the sequence.

2.4 Framework Model

Visualization systems are characterised by an interactive procedure initiated by the user, in which the user keeps interacting as the visualization system feeds results back. This leads to a model of the visualization system which, like the Watson model of which it forms a part, is cyclic. Interaction by the user, initially in terms of the user interface with which he is presented, is logically transformed back into the terms of the science under investigation and updated results are produced by the visualization system and fed back to the user. By this means the user gains insight into the validity of his original hypothesis and the model by which he is investigating it.

As a refinement of this model, we have defined the visualization system in terms of a set of general purpose abstract modules which may be combined in various ways to provide the application required. It should be stressed that this is a logical model of the visualization system. It does not imply that any such system is has to be implemented in water-tight modules that exactly mirror the modules of the model.

We suggest that visualization can be considered a window into the physical parts of the scientific process, and as such it should be able to be placed anywhere to get an understanding of any part of the cycle. It should be emphasised that visualization is only one of a number of tools which could be used for this purpose. As such, it has its own strengths and weaknesses which make it more or less appropriate for various jobs. Visualization is not simply the generation of final images, but should be considered as part of the scientific toolkit.

Given that the cyclic model is a useful abstraction, where and how might visualization be useful? It is obvious that visualization of simulation results, or of experimental or observational data, is useful and may be an end in itself. What is not so obvious is that visualization may play a large part in forming the link between hypothesis and experiment, or between insight and new hypothesis.

2.5 Module Model

2.5.1 *Overview*

The Visualization System is modelled as a controlled dual dataflow into which modules can be inserted to form the required Visualization Tool. The modules defined permit the user to construct a tool tailored to his specific application.

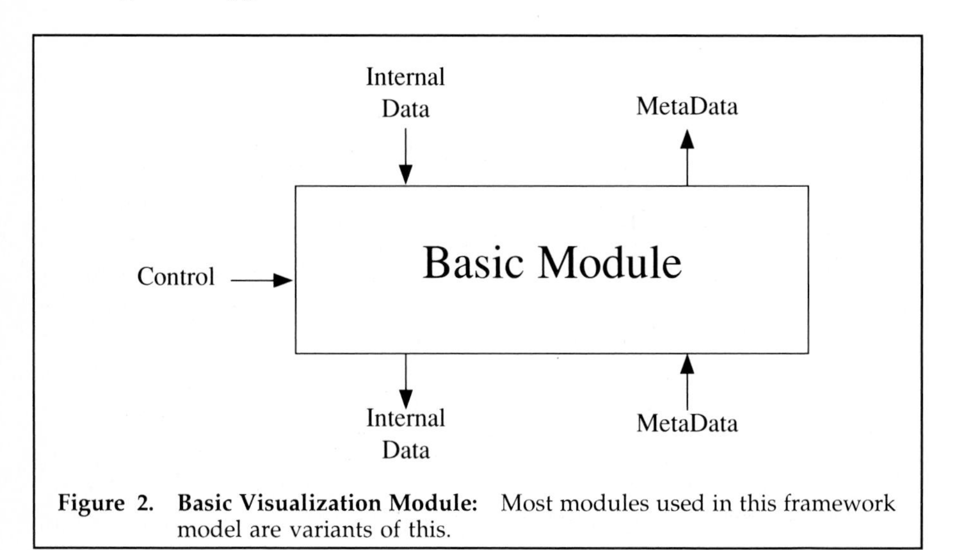

Figure 2. Basic Visualization Module: Most modules used in this framework model are variants of this.

The basic module intercepts both a data flow towards the user (output) and a data flow away from the user (input) and responds to controls from the Command Interpreter. The dual dataflow is terminated at one end by a Command Sequencer module which is responsible for converting input into a command sequence. At the other end the dataflow is, for completeness, terminated by a "user module" (that is, the user!). The user may be considered to be the agent that converts the output of the visualization system into further command input.

Although many simple visualization systems may be constructed as a linear sequence of modules, the model does not forbid forking and merging of the data paths. Multiple users (or a single user with multiple User Interfaces) can be accommodated, as can multiple command sequencers and generators.

The model does not specify how the data paths should be implemented. Although the three paths (control, input, output) could each be implemented as a bus, this is not a feature of the model since it might constrain the parallel execution of the various modules in a system.

The simple example of Figure 3 on page 22 shows a useful minimal configuration of modules that might be used as a system for drawing contour maps from raw data. The linear sequence of modules

- Data Manipulation
- Visualization Technique
- Base Graphics System
- User Interface

will be a feature of many constructed systems although Data Access (or Import/Export) modules may be interposed. Similarly the Command Interpreter and Command Sequencer will rarely, if ever, be separated other than to interpose a Data Access module for macro archive purposes. In the system depicted the primary data source is external, the data being captured by an Import module into the "top" of the (output) data flow. The captured data may be stored internally by the Data Access module and may subsequently be "replayed" into the system instead of (or as well as) the primary external data source. In this simple example, Command Data returns through each module in turn, allowing each module to apply an appropriate reverse mapping. A possible alternative might have been for Command Data to have returned directly to the Command Sequencer from the Base Graphics System.

2.5.2 *Details of Data Types*

The model employs a small set of data types to describe the dataflow between the logical modules.

Control Data
is parametric coordination information. It activates and controls the modules in the system. It is generated only by the Command Interpreter and is implicitly connected to all control inputs.

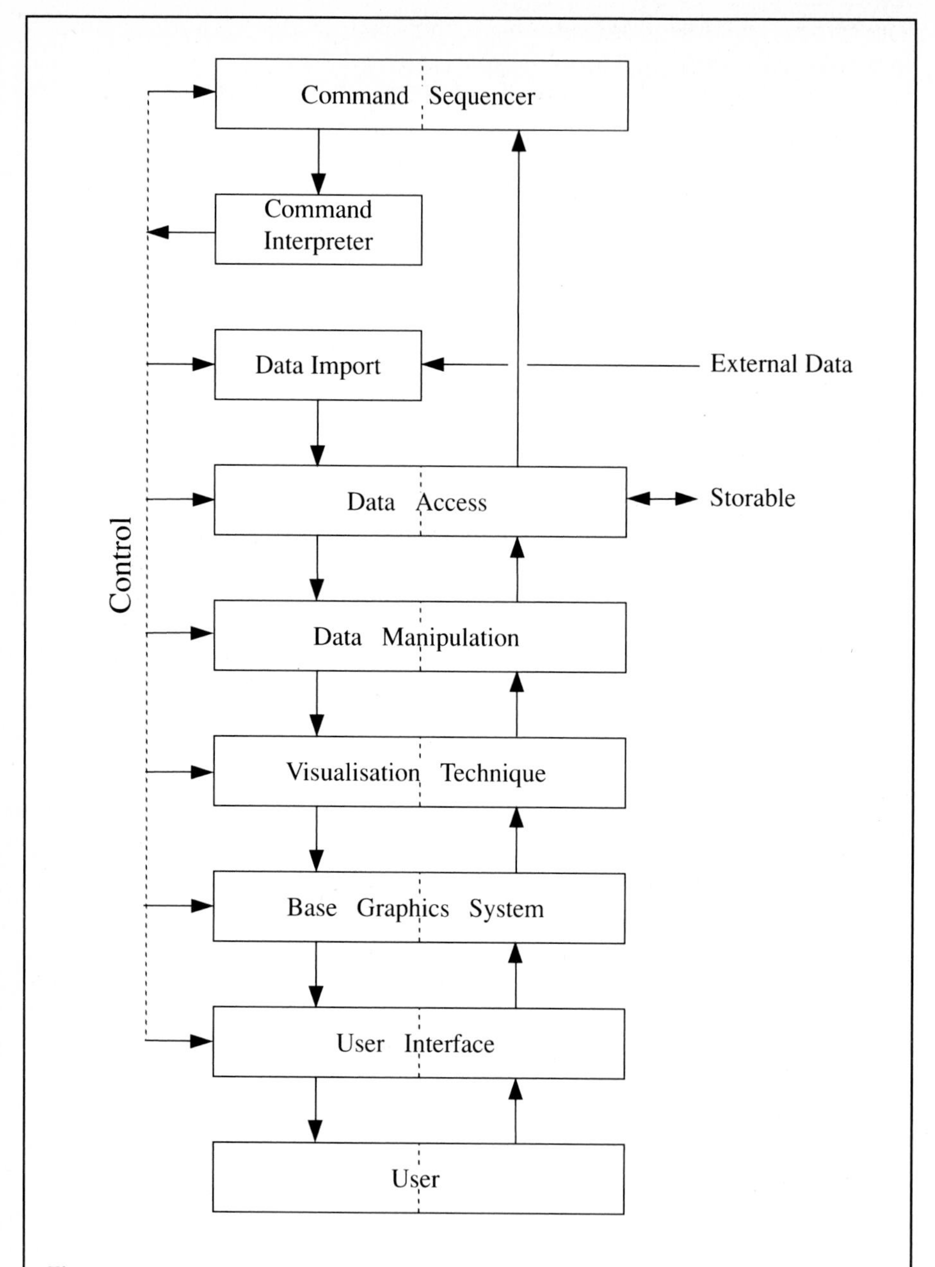

Figure 3. Model of a Graphics System Drawing a Contour Map: This shows how the framework modules and data types can be used to model a very simple graphics system, reading data from an external source and producing a contour map on a screen.

User Input/Visual Output

These two data types encompass all forms of communication between the user and the system, such as a keyboard, mouse, hardcopy device, screen, lightpen.

External Data

may be imported or exported at any level of the Internal Data hierarchy, e.g. observed, experimental or simulated (scientific) model data, CGM, PostScript, TIFF.

Storable Data

is capable of being stored and retrieved within the Visualization System. A Data Access module can convert between Storable Data and all levels of the Internal Data hierarchy (including Command Data in the (input) data flow from the user).

Internal Data

is the general term for all information permitted to pass along the dual flow of the model (excepting the Visual Output and User Input links). At some links in the data flow, only reduced forms of Internal Data are permitted.

Graphics Data

is a reduced form of Internal Data, encompassing geometric and property data, derived by a Visualization Technique. Typically, this would represent graphics primitives.

Picture Data

is a reduced form of Graphics Data derived by a Base Graphics System. Typically this would be a pixel map but this could be primitives for 2D display or hardcopy.

Command Data

is a form of Internal Data, possibly incompletely specified, passing from the User Interface towards the Command Interpreter (by which point it will have become completely specified). Command Data, which may include values, is transformed and made more specific as its passes through transformation modules.

The many forms of external data and ways in which data may be stored and organized are examined in detail in chapter 4 on Data Facilities.

2.5.3 *Details of Modules*

The following sections describe the outline form of each identified module type. In a real system there would be a library of modules of each type and in many cases the facility for the user to construct or install new modules within these types (for example to define a new Data Manipulation module for a specific application).

2.5.3.1 *User*

The User is an essential part of every scientific visualization system and so a User Module is defined in this framework. The User captures the Visual Output of a User Interface module and provides the initiating input to command the system.

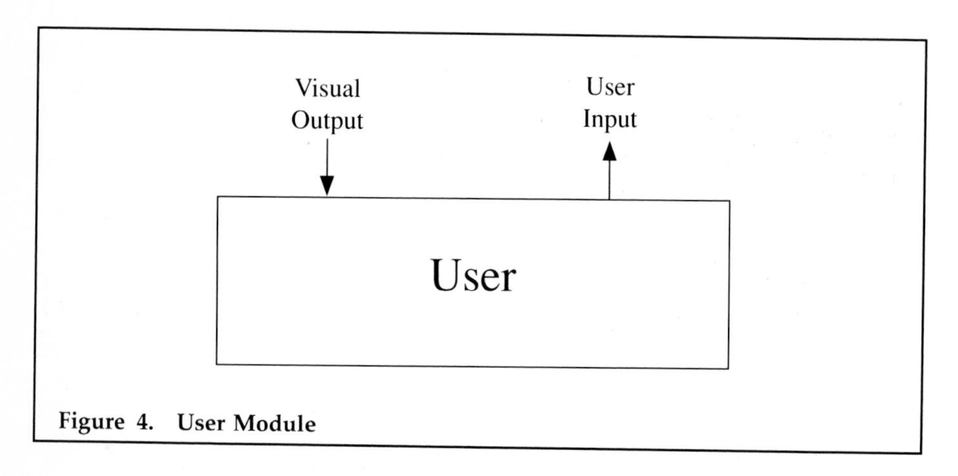

Figure 4. User Module

2.5.3.2 *User Interface Module*

A User Interface module maps between the user's world and the internal data paths of the system. An Application Builder (Network Editor) may be provided in this module - indeed if such a facility is available it is bound to be (entirely contained) in this module since the Builder must be able to function independently of the data paths to other modules.

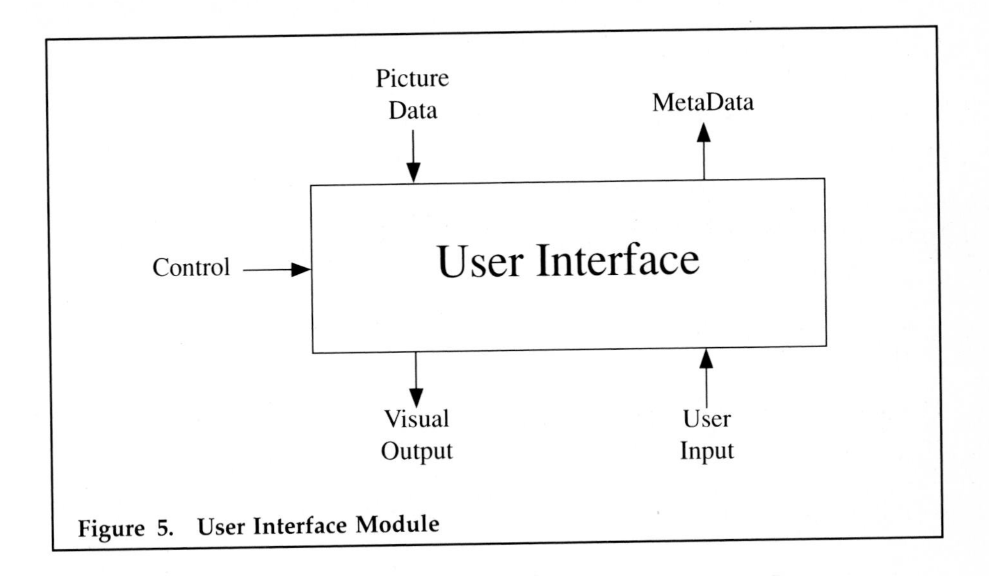

Figure 5. User Interface Module

2.5.3.3 *Base Graphics System*

A Base Graphics System module reduces geometric (and property) data to the form required by a User Interface module. Examples include graphics standards such as GKS, GKS-3D, PHIGS and commercial products such as GDDM, PLOT-10 and GINO-F amongst hundreds of others.

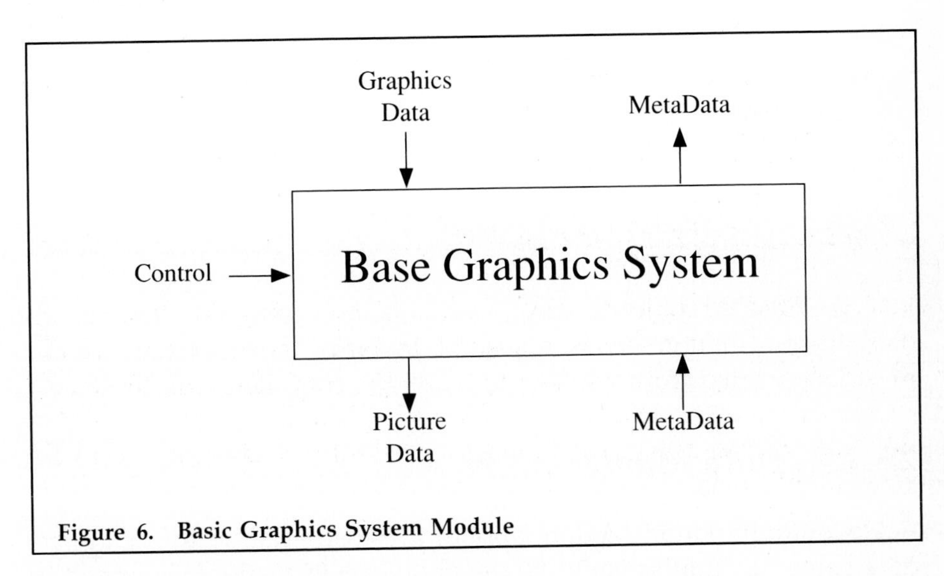

Figure 6. Basic Graphics System Module

A Base Graphics System module would normally be expected to include the functionality for import and export of graphics metafile information. If the Base Graphics System did not provide this, the model allows for the functionality to be described by means of Data Import and Data Export modules sandwiching the Base Graphics System.

The Base Graphics System may transform not only data headed towards the user but also metadata headed from the user towards the command interpreter.

2.5.3.4 *Visualization Technique*

A Visualization Technique transforms data derived from the scientific model to that required by a Base Graphics System module. Numerous examples of Visualization Techniques are presented in chapter 3.

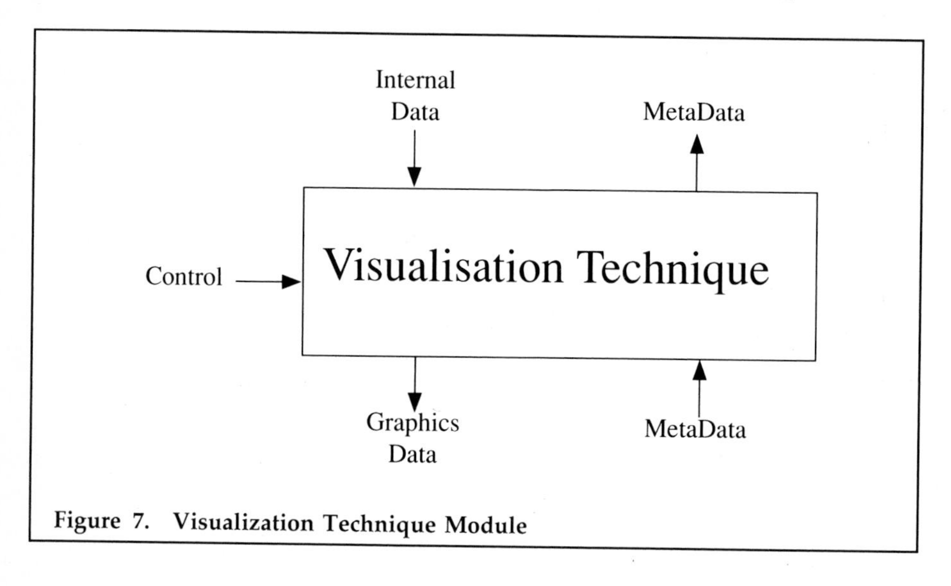

Figure 7. Visualization Technique Module

Like the Base Graphics System, a visualization technique implies not only transformation of data headed towards the user but also of metadata headed from the user towards the command interpreter.

2.5.3.5 *Data Manipulation*

A Data Manipulation module applies a transformation to each of its data paths. Examples could be the data transformations described in chapter 4.

Like the Base Graphics System and Visualization Technique modules, a data transformation implies transformation of metadata headed from the user towards the command interpreter.

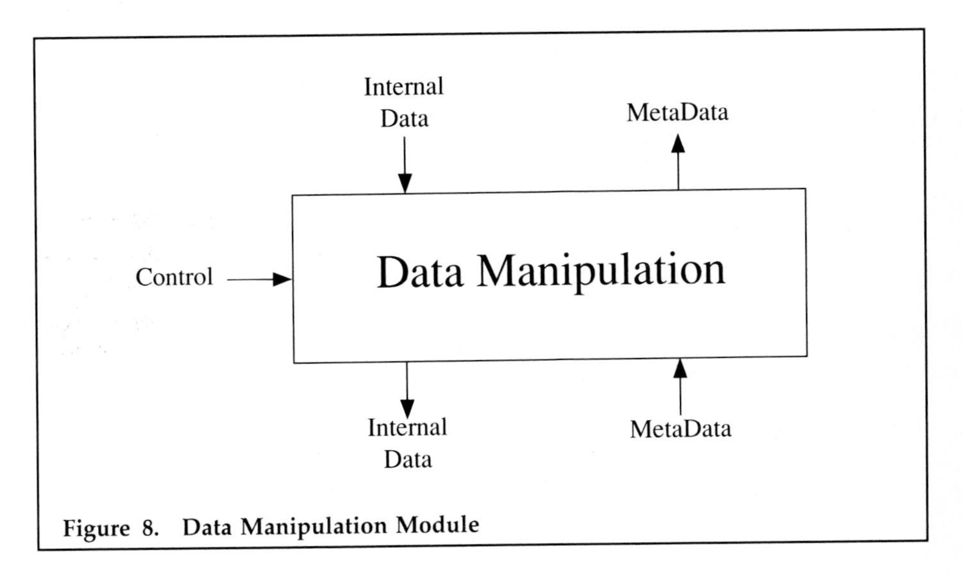

Figure 8. Data Manipulation Module

2.5.3.6 *Data Access*

A Data Access module allows information passing along the data paths to be stored and stored data to be retrieved into either data path.

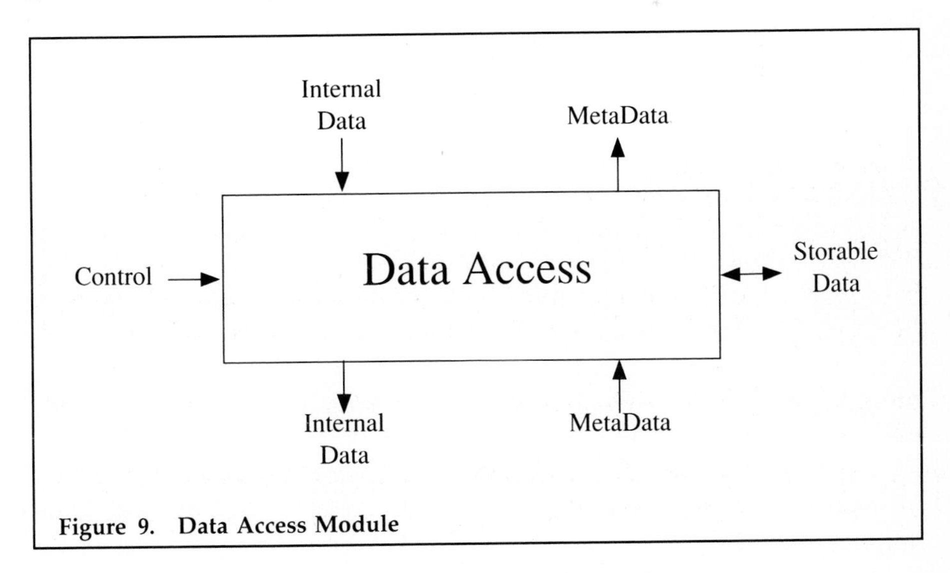

Figure 9. Data Access Module

Except perhaps for tagging to indicate that data has been stored, the data flows in both directions are propagated without modification. In systems constructed to visualize previously captured data, a Data Access module is very likely to be configured with no Internal Data input.

2.5.3.7 *Data Import*

A Data Import module allows External Data to be captured into the (output) data flow. Although the Data Import module is constructed so as to import data into an Internal Data path that is otherwise propagated unmodified, many systems will use this module as the primary data source in which case it is likely that the module will have no Internal Data input.

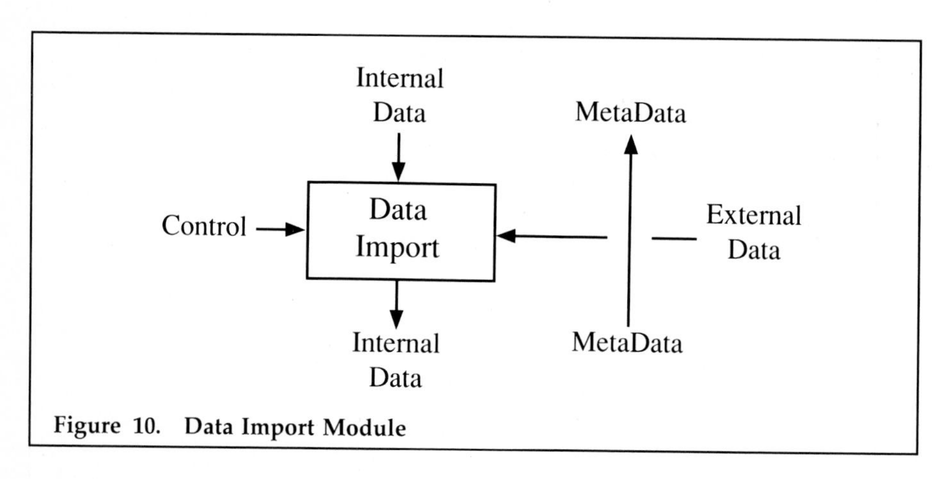

Figure 10. Data Import Module

2.5.3.8 *Data Export*

A Data Export module allows information passing along the (output) data path to pass out of the Visualization System. The module is also expected to propagate the Internal Data onwards without modification (except perhaps to tag the fact that is has been exported). It is, however, possible to construct a system in which there is no other module for the Data Export module to propagate the data to (for example where Internal Data has been forked into a Data Manipulation module to prepare it for the Data Export module). Note that in such a case there would be an unused (return) Command Data path.

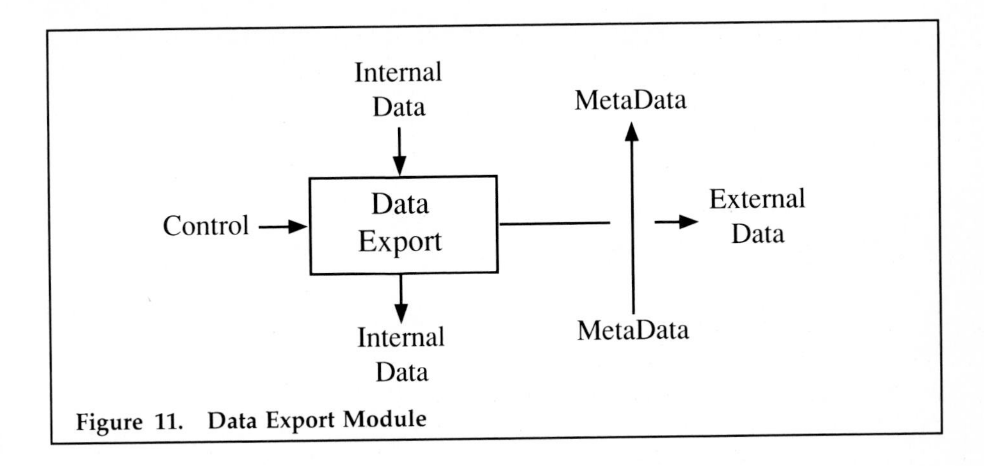

Figure 11. Data Export Module

2.5.3.9 *Command Interpreter*

A Command Interpreter module translates the (fully defined) Command Data into Control Data for the other modules in the Visualization System. Note that, unlike all other modules, the Command Interpreter does not propagate any data directly into the (output) dataflow; any associated data required by Control Data is passed directly to the module concerned along the Control path.

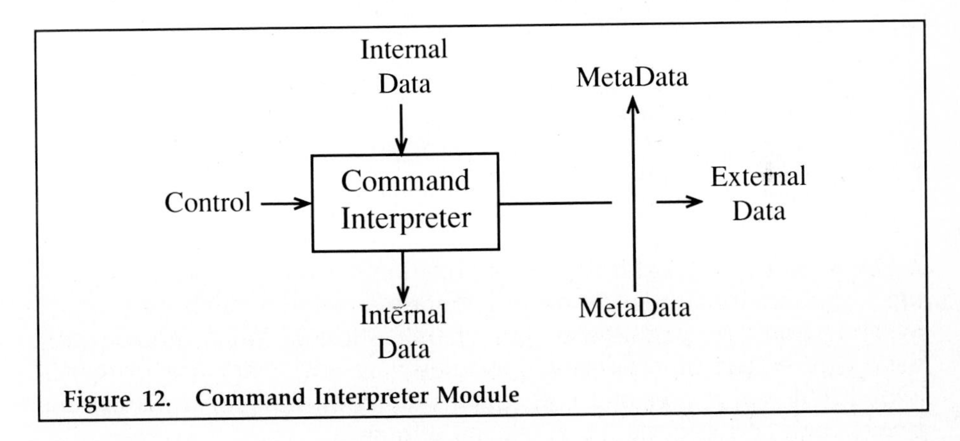

Figure 12. Command Interpreter Module

2.5.3.10 *Command Sequencer*

A Command Sequencer module provides the final opportunity to complete the definition of the Command Data. The Command Interpreter function is separated out from the Command Sequencer so that a Data Access module can be configured between them. This arrangement provides the storage and retrieval that might be required to support the provision of macro and animation facilities by this module.

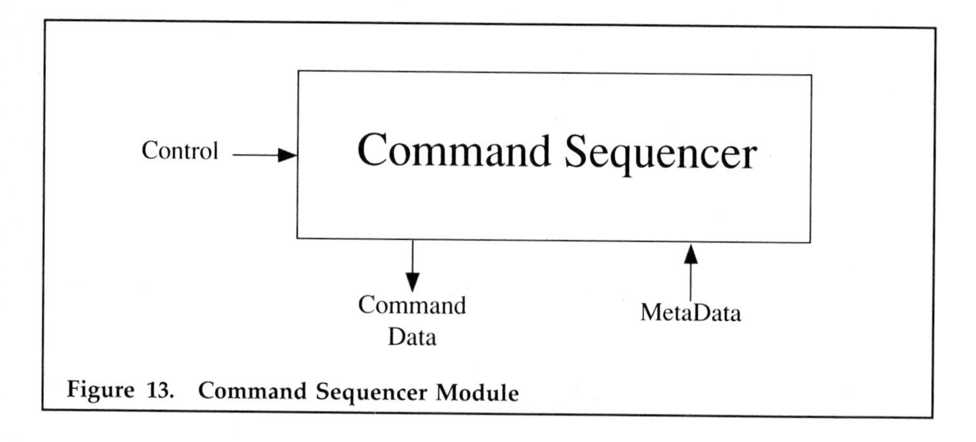

Figure 13. Command Sequencer Module

2.6 Functionality Aspects

While the functional capabilities of a visualization system may be described in terms of the "module model", this does not indicate how rich the visualization system may be in terms of functionality. This section describes the functional aspects of the system that are not quantified by the "module model".

2.6.1 *Functional Richness*

For each type of module, if a visualization system provides that function, it may provide a number of different options.

As an example, consider the "Visualization Technique" class of module. All visualization systems will contain a range of such techniques. Systems that are targetted at the visualization of solid models may have a selection of rendering modules. Systems that are primarily aimed at image processing will contain many options for transforming image data and generating displayable images.

2.6.2 *Data Import and Export*

The usefulness of a visualization system may depend critically upon the range of data import and export functions available. With scientific disciplines defining their own data formats at an alarming rate, the most likely solution to this problem is for any visualization system to have a flexible approach to the addition of new data import and export modules.

It should also be noted that encouragement by national, international and discipline-specific bodies towards the use of existing standards in this area (and the reduction in the number of such standards) would avoid great wastage of effort.

2.6.3 *Data Accuracy and Errors*

There are two ways in which data accuracy and errors may be a critical aspect in a visualization system.

- For experimental data that may contain errors, it may be important to preserve these errors, handle them fully in the internal data system and provide means of manipulating and displaying them.
- For all of the processes within the visualization system, it may be useful to be able to estimate and even display errors that are created by the visualization process itself.

External data may have noise or similar errors. The "Data Import" modules within the visualization system may filter the data as it is brought in (so that the internal data does not carry error information in it) or may import the error information as well. Similarly the "Data Manipulation" and "Visualization Technique" modules may provide for treatment of errors contained in data and their explicit display.

Another major concern to users is the introduction of errors by the visualization process. There is a need for a systematic approach to error indication and control. The question of warning the user about the appropriateness of an action and the likelihood of error introduction needs to be addressed.

2.6.4 *Presentation of Metric Information*

When visualizing scientific information, there will be times when a rendering of the data (for instance, a perspective view of a terrain) is required, times when the value of the data is required (in the same example, this might be a spot height) and times when both need to be seen simultaneously.

Thus there may be a requirement, according to the application, for one or more forms of numerical annotation:

- display of scale information on axes;
- superimposition of calibrated scales (such as contours) on the basic display;
- feedback of cursor information transformed back to the scientific system's coordinate system;
- indication of the form of coordinate system (for example linear, logarithmic, exponential, polar) is use;
- more complex displays of numerical values based on the displayed and/or selected data.

It is also important to have a readily understandable scale setting process. The system must be developed in such a way that it is always able to give useful information to the user. This information must be sensibly combined with the visualization output.

2.6.5 *Distributed Architecture*

A distributed architecture may be appropriate for a scientific visualization system for a number of reasons.

- In many cases the data of interest to the scientist is massive and stored on another computer system. The scientist immediately becomes involved in the mechanisms for access to, and selection from, large external data stores.
- Many problems that fully warrant the use of a visualization system are attacked by teams of scientists remote from each other, and so the sharing of both data and results becomes an essential tool required for the team to make progress.
- The computer resources directly available to the scientist may be insufficient for aspects of the visualization process and so other, external, computer power is often brought in to deal with some aspects of the research.

In terms of the "module model" described above, the details of the implementation of any module (or combination of modules) is not specified, so any may be implemented by means of a distributed architecture. For the user of the system, the extent to which this method of implementation is transparent in terms of function and performance may be crucial.

2.7 Qualitative Aspects

There are other aspects of the behaviour of a visualization system that are not reflected by these models.

These qualitative aspects

- cannot be modelled,
- are difficult to quantify, and
- are not subject to any strict recommendations or advice.

2.7.1 *Responsiveness*

This stresses the importance of having an appropriate response time from the visualization system. This response time can be considered as the time taken to complete a circuit around the "module model". In more general terms it may also be applied to a series of interactions with the visualization system.

In some cases a prompt visualization response cannot be achieved with the present level of technology. In addition, users' needs are growing faster than technology progresses. However, there are some cases where compromise is possible: for instance, it may be more acceptable to produce a fast but coarse representation than a slow but finely rendered one. This compromise introduces a whole range of new questions that are of a more or less ethical or moral nature. For example, in life-threatening domains such as surgical and radiotherapy treatment planning, should one treat fewer people accurately, or more people less accurately?

2.7.2 *Human-Computer Interface*

The way in which the user and the visualization system interact is highly complex, involving cognitive and perceptual issues as well as more mechanical (system-defined) aspects. Different disciplines have developed systems - and hence interfaces between users and computers - to suit their needs and match their budgets. While there are an increasing number of shared technologies and methods of working, many issues still deserve examination and this is done in chapter 5 which is devoted to the Human-Computer Interface.

2.7.3 *Appropriateness*

As in virtually every aspect of computing, any system may be evaluated as adequate or inadequate only when measured against the requirements of a particular application. A system designed for image processing should not be considered a poor system if it is incapable of volume rendering! Each visualization system will, explicitly or implicitly, have a defined scope and an anticipated area of application.

2.7.4 *Costs versus Benefits*

Visualization systems and the hardware platforms on which they are run cost money. Use of a visualization system therefore has a quantifiable cost. An appropriate visualization system should improve the speed with which the scientific work can progress and this implies a benefit to the user and the research team.

In each area it should be possible to produce an analysis of the relative costs and benefits of using the visualization system. In extreme cases, the scientific work may not be possible at all unless a visualization system is used.

2.8 Implementation

In general the way in which a visualization system is implemented is not relevant: what matters is the end result as seen by the user. However, the complexity of the structure of visualization systems has encouraged the use of "visual programming techniques" where the user assembles a visualization system from a library of modules. This was the method adopted for AVS [Upson89a] and a similar approach was adopted in ConMan [Haeberli88]. While these systems assume a network structure for the visualization process, the module model presented in this chapter may be seen as functionally equivalent.

2.9 Key References

Most references are provided in the bibliography at the end of the book. A few references which are considered key for this chapter are provided here.

[Haber90]

Haber R.B., McNabb D.A., "Visualization Idioms : A Conceptual Model for Scientific Visualization Systems", *Visualization in Scientific Computing,* ed. Nielson G.M., Shriver B., Rosenblum L.J., **(1990)**, *pp 74-93.*

This provides a deep analysis of the parts of a visualization system that may be computerized, concentrating on the transformations from "simulation data" (which includes data generated by analysis) to "displayable image".

[Wallas26]

Wallas G., "The Art of Thought", **(Jonathan Cape, London, 1926)**.

Subsequent psychological studies on creative thought refer to the four stages in creative thinking described by Wallas.

[Dyer90]

Dyer D.S., "A Dataflow Toolkit for Visualization", *IEEE Computer Graphics and Applications* vol 10 (4), **(July 1990)**, *pp 60-69.*

This article about apE 2.0 explains some of its design ideas and these may be seen as an implementation of some aspects of the module model presented in this chapter.

[Rasure91]

Rasure J., Argiro D., Sauer T., Williams C., "A Visual Language and Software Development Environment for Image Processing", *International Journal of Imaging Systems and Technology,* **(1991)**.

This is an article on Khoros, a visualization system again based on a modular scheme. language.

[vandeWettering90]

van de Wettering M., "The Application Visualization System - AVS 2.0", *Pixel,* **(July / Aug 1990)**.

This article describes AVS, perhaps the most well known modular visualization system.

Chapter 3

VISUALIZATION TECHNIQUES

Edited by Ken Brodlie

3.1 Introduction

The previous chapter identified the concept of a Visualization Technique module which is responsible for generating and manipulating a graphic representation from a set of data, allowing investigation through user interaction. The purpose of this chapter is to describe this concept in some detail.

As is evident already from this book, visualization in science and engineering properly encompasses much more than graphic representation. It involves gaining insight and understanding of the problem solving process. Thus visualization is inherently application-dependent, and many techniques only make sense within a particular context. Some of these more application-specific techniques are discussed in the later chapter on Applications.

However there are a range of techniques that are generic in nature, and can be tailored to different applications. It is these generic visualization techniques which we describe in this chapter.

There is an important point to make at the outset. The data which is fed to a visualization technique is typically sampled from some underlying physical phenomenon. It is this underlying phenomenon that we are aiming to visualize and hence understand - not the data itself. This distinction is fundamental.

There are three distinct parts to a visualization technique:

- the construction of an empirical model from the data, that can act as a representation of the physical phenomenon;
- the selection of some schematic means of depicting the model;
- the rendering of the image on a graphics display.

These steps define the basic structure of a visualization technique, and are discussed in more detail in the next section.

It is impossible in this chapter to cover every visualization technique of general interest - there are simply too many. Also we make no attempt to be prescriptive or to recommend particular techniques for particular situations. Rather the chapter concentrates on establishing a means of classifying techniques, together with a variety of examples. We show how these examples fit into the classification system, so further examples can be added. As a result, we include simple visualization techniques, such as line graphs, as well as more complicated techniques, such as vector fields and volume visualization. The aim is to see all techniques as members of a family.

The classification is based on the dimensionality of the entity the particular technique is depicting. The scheme is described in section "Classification" on page 40. The different techniques are themselves described in sections "Techniques for Point Data" on page 43 to "Image Processing Techniques" on page 66.

A number of additional sections describe related areas. One section deals with image processing techniques, which allow manipulation and investigation of images created within a visualization system. Another section deals with animation, where time is a parameter of the entity being visualized. A section looks at interaction, and the different techniques for interrogating a picture. Finally a section looks at aids to perception.

3.2 Elements of a Visualization Technique

This section describes the three main elements of a visualization technique. The model we use follows closely that proposed in [Haber90] and we use their terminology where possible.

3.2.1 Building an Empirical Model

In this operation, an internal model of the physical entity is constructed from the data provided. For example, in the case of contouring from height data given at a set of scattered points, a continuous function interpolating the data is constructed. Some associated data may be supplied to assist in this - for example, a Delaunay triangulation of the data points, or information that the entity has some particular property, such as being everywhere positive.

This reconstruction step helps us to distinguish the dependent and independent variables in the model. In the contouring example, the space variables are independent; the height variable is dependent on the two space variables.

It is interesting to note that this step involves different aspects of mathematics. Sampling theory provides the conditions under which a continuous signal may be reconstructed from a set of samples: essentially, the original signal must be sampled at a frequency greater than twice the highest frequency component in its spectrum - this lower bound being known as the Nyquist frequency (see, for example, [Foley90b]). Numerical analysis provides a variety of interpolation algorithms for constructing continuous functions through sampled data.

It is often the case that the data is known to be in error. The use of interpolation may then be inappropriate; it is wrong to look for a model that matches every data value, for then errors are tracked rather than the underlying trend. Instead of interpolation, an approximation process is required, in which the model is not constrained to satisfy each data value. An iterative process may be required to find the most appropriate of a class of models - for example, in curve fitting, the position of the knots in a cubic spline may be varied until an error criterion is minimized.

3.2.2 *Representation as an Abstract Visualization Object*

In this operation, the empirical model is represented as some abstract visualization object. In the example above, this means associating the model with, say, a shaded contour map - the shading indicating regions between two contour levels. The contour levels can be regarded as parameters of the visualization object. Another abstraction could have been used - for example, a surface view.

This stage could be considered as the scientific stage of the process. The choice of abstraction is made so as to learn most about the underlying phenomenon.

3.2.3 *Realization of the Visualization Object*

In this operation, the abstract visualization object is realized as a graphics sequence. In the contouring example, appropriate "Fill

Area" primitives are generated, together with attributes (interior style, colour, etc.) specifying how the areas should be rendered.

In essence, this final step can be regarded as the engineering element. The view for the scientist is constructed on the graphics display surface.

Thus the study of visualization techniques encompasses the three disciplines of Mathematics, Science and Engineering. Several processes are involved in the transformation from data to graphics display. Though the result aims to help the scientist understand his data, a word of caution is also in order. The transformation steps may introduce error or artefacts not present in the data, and great care is always needed. The paper by Buning [Buning88] is salutary reading.

3.3 Classification

Suppose that an empirical model has been created from the supplied data, giving some entity to visualize. This entity will have a range of values, and will typically be defined over some domain of independent variables. Mathematically we can express an entity therefore as a function F of many variables:

$$F(X) \quad \text{where} \quad X = (x_1, x_2, \ldots x_n)$$

The dimension of the domain is n.

The function F can be one of a number of different types: a scalar, a vector ($F = (f_1, f_2, \ldots f_k)$), or even a tensor of second order ($F = (f_{jk})$) or higher. The classification of techniques will be based on the type of the function, and the dimension n of the domain. There is an additional case where the requirement is simply to visualize a set of points in the domain (the familiar scatter plot), in which case there is no associated function, just a set of points.

It is helpful to introduce some notation at this stage. A scalar entity S with domain of dimension n will be identified as E_n^S; superscripts V and T will represent vector and tensor entities respectively. The dimension of the vector can be added as a suffix (for instance V_3), with tensor dimensions handled similarly. Thus $E_n^{V_5}$ denotes a vector of length five defined over an n-dimensional domain. A set of points will be denoted as E_n^P.

This separation into point, scalar, vector and tensor fields is the major classification we shall use. The domain dimension provides a subclassification, and indeed we make further distinctions according to the nature of the domain. Three cases are identified:

1. The entity is defined pointwise over a continuous domain - for example, in a terrain map, the height is defined separately for each point.

2. The entity is defined over regions of a continuous domain - for example, a chloropleth map showing the density of population in each country is of this type. To indicate this we use the notation $E_{[2]}$ to indicate a two-dimensional domain, with the entity defined over regions rather than at individual points.

3. The entity is defined over an enumerated set - for example, a chart showing the number of cars of each manufacturer sold in a particular year. To indicate this we use the notation $E_{\{1\}}$ to indicate a one-dimensional domain consisting of a set of enumerated values, i.e. car manufacturers.

There is one further extension of the notation to describe. It is quite common, especially with scalar fields, to have an entity that is a set of values. For example, one may wish to visualize the pressure and temperature within a volume, i.e. two scalar fields over the same domain. The requirement is to show both fields on the same graphical representation, so as to understand the relation between the two fields. This will be represented as E_3^{2S} to indicate a set of two scalar values over 3D domain, the general notation being E_n^{mS}.

Some further examples should help make this notation clear. Consider first some scalar fields, beginning with a simple line graph. We are given a set of data points, the y-values being samples of $F(x_1)$ at a set of points. The entity to be visualized is the curve $F(x_1)$ - of type E_1^S.

Now consider volume visualization, in which the data supplied are density values at points on a regular grid within a cuboid. The entity to be visualized is a scalar function $F(x_1, x_2, x_3)$ underlying the data. This is of type E_3^S.

A 2D scatter plot is classified as E_2^P. The generalisation to multivariate scatter plots is obvious - they are of type E_n^P.

Another major class is vector display techniques, where the entity being displayed is a vector. An example is the visualization of a 3D vector field in a volume. This is of type $E_3^{V_3}$, i.e. 3D vector over 3D domain.

Thus the notation covers a very wide range of visualization techniques. Indeed the concept can extend nicely to more complicated examples. Suppose we wish to visualize the age distribution of the population in Scotland and England. The domain is the set of the two areas representing the countries - this would be of type $E_{[2]}$. However the range is another entity - namely, a set of histograms. A histogram can be regarded as of type $E^1_{[1]}$ - a scalar function (number of people) defined over regions (age ranges) in a one-dimensional domain. Thus the composite representation can be identified as the type $E_{[2]}^{E^1_{[1]}}$!

Other researchers have created different classifications. For example, Bergeron and Grinstein [Bergeron89] introduce the concept of lattices to describe different arrangements of data. The notation L_q^p indicates a p-dimensional lattice of q-dimensional data. The dimension of the lattice indicates the ordering of the data: zero-dimensional is unordered (for example a set of points), one-dimensional indicates a vector of data elements (for example a list of points to be connected by a polyline) and two-dimensional indicates an array of data elements (for example height and pressure at nodes of a regular grid). This classification is more oriented to the data, rather than the entity underlying the data. For example, in a contouring example, Bergeron and Grinstein would distinguish data collected at random, from data collected on a regular grid; our scheme would not, because the underlying entity is the same. On the other hand, our scheme distinguishes a 2D vector from a pair of unconnected scalar values, and makes a clear distinction between dependent and independent variables.

One special comment is needed. Frequently an independent variable X contains space and time components; indeed time is a sufficiently special case to be separated out. Thus an entity which varies with time is modelled as $F(x_1, x_2, \ldots x_n; t)$. Time-dependent visualization can be naturally handled by animation: a sequence of images with successive time stamps are generated on video for subsequent display. Animation is described separately in section "Animation" on page 69. If animation is not available, static images can be produced simply by taking time as an extra dimension - using a technique of type $E_{(n+1)}$.

A final word on notation. We have used the superscript and subscript notation to reflect that the entity is defined ON the domain, and YIELDS a scalar, vector or tensor. An alternative notation which some may prefer, is to use a functional approach: $E_3^{V_2}$ being written as $E(V{:}2,3)$. The composite map and histogram example above becomes $E(E(1,[1]),[2])$.

Dimen-sion	Point	Scalar	Vector	
			2D	3D
1	1D Scatter Plot	Line Graph		
[1]		Histogram		
{1}		Bar Chart		
2	2D Scatter Plot	Contour Plot Surface View Image Display	2D Arrows 2D Stream-lines 2D Particle Tracks 2D Field Topology	3D Arrows in Plane
[2]		Bounded Region Plot 2D Histogram		
{2}		2D Bar Chart		
3	3D Scatter Plot	Isosurfaces Basket Weave Volume Rendering		3D Arrows in Volume
n	n-dimen-sional Scatter Plot			

Figure 14. Summary of Visualization Techniques: This table shows where the various visualization techniques fit into the classification scheme used in this book.

In the following sections, the techniques shown in the above table are described in more detail.

3.4 Techniques for Point Data

This category includes the major area of visualization of multivariate data, where each element of data is considered as a point in the multi-dimensional space. The analysis of such data can be greatly assisted by the use of visual techniques: the challenge is to somehow project the data from the multi-dimensional space to the 2D viewing surface.

One-Dimensional Scatter Plot

Category: E_1^P

This is the simple case, where the values may be marked as points on a single axis. An example could be the distance of other planets from the Earth.

Two-Dimensional Scatter Plot

Category: E_2^P

This is the traditional 2D scatter plot, where the pairs of values are represented as points in the plane. An example is the height and weight of a set of individuals.

Three-Dimensional Scatter Plot

Category: E_3^P

For the 3D case, it is possible to project the points to a 2D plane and take some attribute of the marker glyph used to indicate a third component. Colour can be used, or the size or type of glyph. However it is generally considered [Crawford90] that motion has the best result, with the points drawn in 3D space and using rotation of the point "cloud" about coordinate axes to help in visualization of the data.

Higher Dimensional Scatter Plot

Category: E_n^P

A number of ideas have been suggested for higher values of n. These include Chernoff faces - see [Chernoff73] and [Pickover90] - where the different variables are "attached" to different features of a schematic representation of a human face: the shape of the eyes, the shape of the mouth and so on. Up to 12 parameters have been successfully represented in this way. Although faces are the most commonly used representation, in principle any suitable real world object can be used; the requirement is that it should have a set of distinct features that can take easily differentiated values - a human body, a car, even a cathedral could be used.

Another useful technique is the use of Andrews plots [Andrews72]. Each of the n-values in a data element ($F1, F2, F3, \ldots$) defines a function:

$$G(t) = \frac{F1}{\sqrt{2}} + F2\sin(t) + F3\cos(t) + F4\sin(2t) + F5\cos(2t) + \cdots$$

which is plotted over the range $-\pi$ to π. Thus an Andrews plot consists of a set of curves, one for each data element. Clusters of elements map to similar shaped curves and are thus easily identified.

Other methods are reviewed in [Crawford90] and [Everitt78].

Note: Histogram techniques are often used in conjunction with techniques in this category, though the entity they represent is of a different category. The essential idea of a histogram is to group members of a set into different subsets, and indicate the number in each subset. Taking the example of the 1D scatter plot above, the corresponding histogram would bin planets within different distance ranges from the Earth. We can regard this as a data manipulation operation. A different entity is being displayed in the two cases. A 1D histogram is an entity of type $E_{[1]}^S$ and is described in the next section.

3.5 Techniques for Scalar Entities

3.5.1 *One-Dimensional Domain*

This is the (apparently) simple case where the data is sampled from a function $F(x_1)$. Three basic techniques are described: a line graph, where the entity is defined pointwise over an interval of the continuous real line; a histogram, where the entity is defined over regions of the real line; and a bar chart, where the entity is defined over an enumerated set.

Finally some notes are included on the case where the sampled values are known to be subject to error, which can occur due to observational errors or inappropriate sampling frequency [Nyquist28], [Shannon49].

Line Graph

Category: E_1^S

Given a set of data points, a line is drawn through the data points to visualize the underlying function $F(x_1)$. An example could be to plot a drug concentration within a patient over a period of time.

Though this is an elementary visualization technique, care is needed in the construction of the empirical model, i.e. construction of the underlying function $F(x_1)$ by interpolation. There is an infinite choice of interpolants. The method used can be linear, or higher order with cubic splines [Lancaster86] being the traditional method. Recent interest has centred on the preservation of shape properties inherent in the data; these shape properties include monotonicity [Fritsch80], positivity and convexity [Brodlie91]. For example, in the case above, it is known that the drug concentration must always be positive and so the interpolant must provide this property.

The data may be accompanied by derivative information (for example, when points are solution of an ordinary differential equation [Brankin89]) and this can be used in the construction of the empirical model.

Multiple Line Graphs

Category: E_1^{mS}

It is often useful to display several graphs on the same plot using different rendering techniques to distinguish the graphs. As well as reducing the space required to present the information, correlations between the variables can be appreciated. An example, showing three variables, is given in Figure 15.

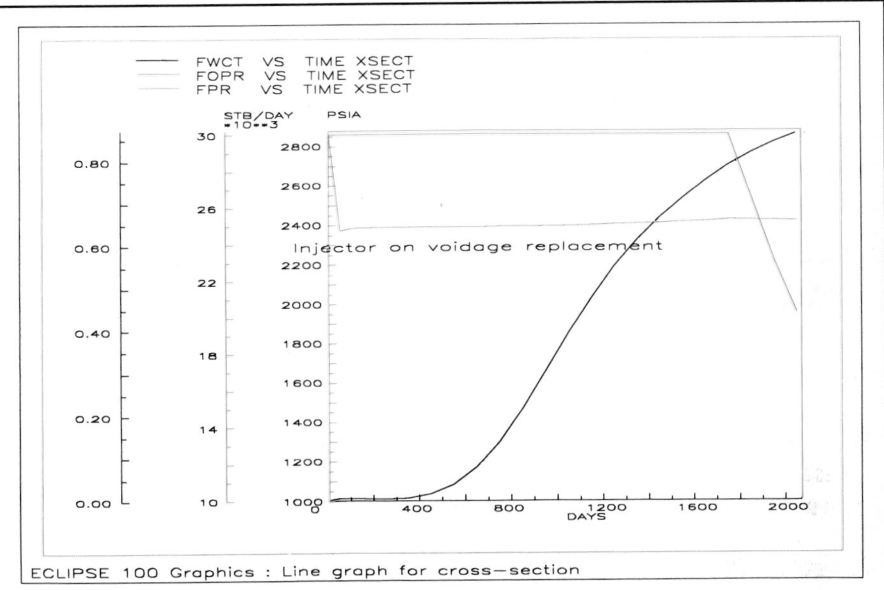

Figure 15. Example of Three Superimposed Line Graphs: The graph shows the change in water production (FWCT), oil rate (FOPR) and oil pressure (FPR) with time over the whole reservoir field. *Produced by INTERA ECL Petroleum Technologies using the Eclipse suite of Oil Reservoir Simulation Software.*

Bar Chart

Category: $E_{\{1\}}^{S}$

Given values of items in a set, a bar chart depicts these values by the length of bars drawn horizontally or vertically. An example is the average cost of a Mars Bar in each year between 1980 and 1990!

There is no mathematical reconstruction here. Often quite sophisticated rendering is done, with shadow effects on bars for example - developments from their frequent use in business graphics.

Note: Pie charts are another technique for this type of entity, when the values can be seen as fractions of a whole.

Histogram

Category: $E^{S}_{[1]}$

Given a set of data values, it is often useful to aggregate the data into bins. A data element is added to a bin if it lies within a corresponding range of values. A histogram shows the number of elements in each bin, by drawing bars of corresponding length.

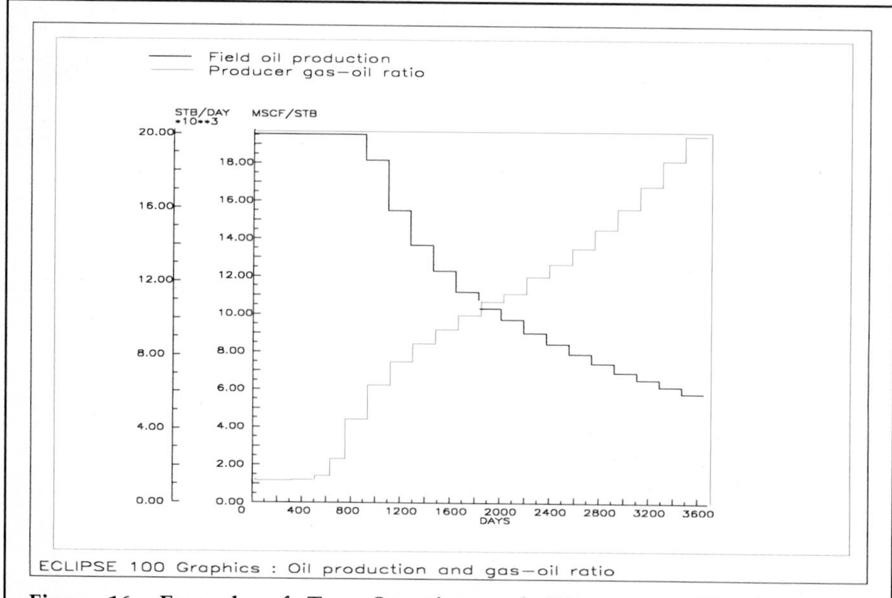

Figure 16. Example of Two Superimposed Histograms: The histograms show the change of oil production for the whole reservoir field and the change of gas/oil ratio for one well, both with respect to time. *Produced by INTERA ECL Petroleum Technologies using the Eclipse suite of Oil Reservoir Simulation Software.*

Multiple values can be displayed by stacking the bars - an entity of type $E^{mS}_{[1]}$. An alternative is to superimpose histograms on the same chart: see Figure 16.

3.5.1.1 *Approximating Functions*

As noted earlier, it is frequently the case that the sampled values are subject to error. The line graph described above, in which an interpolating curve is constructed, no longer reflects the underlying function but incorporates the fluctuations caused by the sample errors. The need is for an approximating function rather than an interpolant.

If the underlying functional form is known, then this can be simply drawn alongside the data points. The errors in the points may be indicated by error bars. It is more usual that the functional form is only partly known, but includes some parameter that may be varied. A minimization procedure is required to establish the best choice of the parameter - best typically in the sense of minimizing the sum of squares of error between data values and approximating curve.

If the underlying functional form is not known, it is common to use cubic spline approximation. Given a set of knots, a spline best fitting the data can be calculated. Varying the knots can lead to an improved approximation. The use of cubic splines for smoothing noisy data is described in [Lancaster86].

A paper by Hopkins [Hopkins90] describes an interactive spline fitting package based on the cubic spline approximation routines in the NAG Library [NAG90].

3.5.2 *Two-Dimensional Domain*

This is the case where data is sampled from a two-dimensional function $F(x_1, x_2)$. There are analogous situations to the one-dimensional case: both bar charts and histograms extend in a natural manner, with types $E^S_{\{2\}}$ and $E^S_{[2]}$ respectively. They are not discussed further here.

We begin with the case where the function is defined pointwise over a continuous 2D domain. An example is the traditional cartographic case of height at a point on the ground. However there are important differences in the empirical model stage depending on how the function is sampled. The sampling may be done on a grid, or at scattered points on the plane.

For gridded data, the simplest interpolation method is bilinear interpolation, relying only on function values at the nodes of the rectangular grid. This model has continuity across the grid lines of value only.

Higher order and higher continuity are achieved by bicubic interpolation, which requires first partial derivatives and mixed second partial derivatives at the nodes. These derivatives may be estimated if they are not given. Shape preservation is sometimes a requirement: Carlson and Fritsch [Carlson85] show how to estimate derivatives so as to preserve monotonicity.

For scattered data, there is a wide variety of methods. Some methods triangulate the points, and construct bilinear or bicubic interpolants over each triangle. For example, Renka and Cline [Renka84] discuss bicubic interpolation over a triangular mesh. Another method is the modified Shephard method, which avoids the triangulation step [Renka88]. Another approach is based on multiquadrics. A more general reference is [Lancaster86].

Again there is an important case where the data are known to be in error. It is quite straightforward to visualize errors in 1D data, for example using error bars, but it is much harder in 2D. A good mathematical method for surface approximation is bicubic spline fitting - [Hayes74] and [NAG90]. The display of the approximant can be handled exactly as for an interpolant. Different methods are now described.

Line Based Contouring

Category: E_2^S

From a set of values at points in 2D plane, isolines of constant value are drawn. The values of the isolines are parameters of the technique. This type of representation was common in the era of vector devices, and remains a useful technique today - for example, in weather forecasting where air pressure is shown by this method in the form of isobars.

The interpolation technique depends on the arrangement of the data points. For gridded data, linear interpolation can be carried out along grid lines to establish intersection points of isolines with the grid; these points are joined by straight lines. Alternatively an interpolant can be created as described above, and isolines tracked step-by-step using successive evaluations of the interpolant.

For scattered data, an interpolant may be constructed and evaluated on a rectangular grid - and then a method for gridded data used.

Alternatively one can use an interpolant whose isolines are easy to draw directly [Powell77].

Derivative information may be available; this can help in construction of the empirical model. Discontinuities may be specified (especially common in geological applications). A good review of contouring methods is given in [Sabin86].

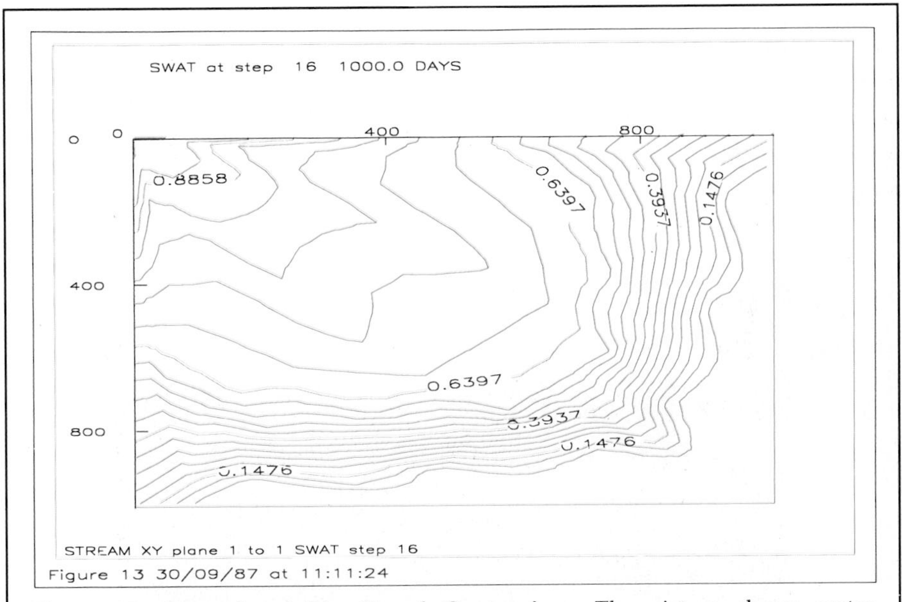

Figure 17. Example of Line-Based Contouring: The picture shows water saturation on a horizontal cross-section of an oil reservoir. *Produced by INTERA ECL Petroleum Technologies using the Eclipse suite of Oil Reservoir Simulation Software.*

Discrete Shaded Contouring

Category: E_2^S

From a set of values at points on the 2D plane, areas between two isolines are indicated by a particular style of shading. A set of isolines is the parameter to the routine. This technique is effective on raster devices.

For gridded data, it is typical to construct a bicubic polynomial in each grid rectangle. For scattered data, it is possible to first interpolate onto a rectangular grid, or to work directly from a triangular mesh - [Preusser86] and [Preusser89].

Image Display

Category: E_2^S

From values at points in 2D plane, an image display of the underlying function is generated. The domain is divided into a grid of cells (typically corresponding to pixels on display), and the colour of each cell chosen to represent the corresponding value of the function.

This technique is often used where there is a dense grid of data (for example from satellites or scanners) and therefore interpolation is rarely an issue. Section "Image Processing Techniques" on page 65 discusses the processing of images.

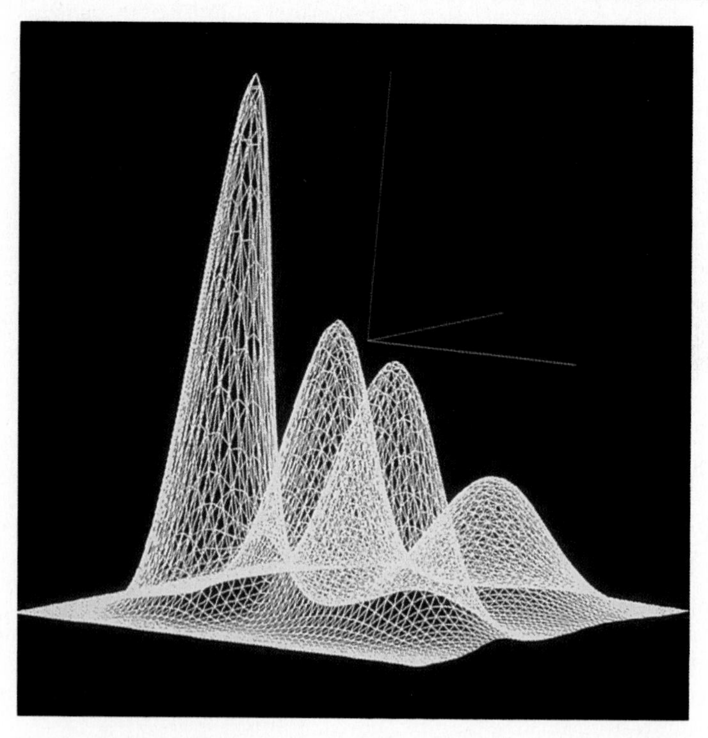

Figure 18. Example of Mesh Surface View: *Produced with Figaro implementation of PHIGS PLUS as part of a student project at University of Leeds by Paul Smith.*

Surface View

Category: E_2^S

A traditional alternative to contouring is to display a mesh of lines parallel to the x and y axes and lying on the underlying surface, the mesh being projected on to a 2D plane. The technique essentially maintains a floating horizon to determine visibility of the surface mesh [Wang90]. The hidden line removal and projection are typically incorporated into the technique, and thus the interface to the base graphics system is 2D. It is possible however to interface directly to a 3D system, delegating the hidden line removal and projection to the base graphics layer. An example is given in Figure 18 on page 52.

Figure 19. Example of Shaded Surface View: *Produced with Figaro implementation of PHIGS PLUS as part of a student project at University of Leeds by Paul Smith.*

Alternatively the surface may be shaded directly, using a standard lighting model, with the mesh optionally overlaid. An example is given in Figure 19 on page 53.

Height-field Plot

Category: E_2^{2S}

This enables two scalar fields over a 2D domain to be displayed. One field is displayed as a surface net, while the other field is displayed as a shaded contour map "draped" over the surface.

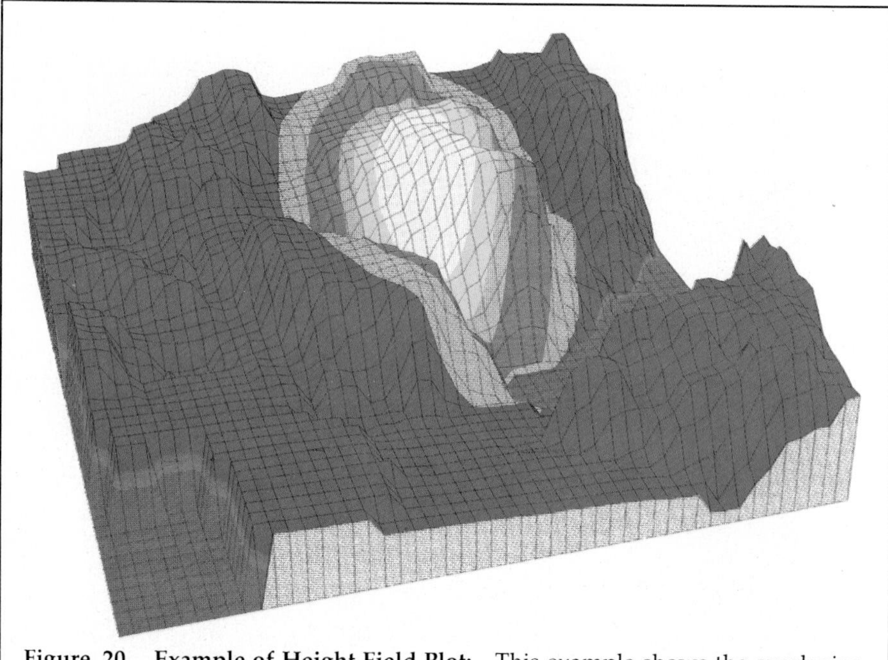

Figure 20. Example of Height Field Plot: This example shows the overlaying of a perspective surface view (here topographic data) with an independent scalar function (here distribution of the mineral Kryptonite in Greenland) which is represented as a shaded contour plot. *Produced with UNIGRAPH 2000 by Malcolm Austen.*

Indeed the other field could be represented as an image display over the surface. An example, presenting topographic data, is shown in Figure 20.

Multiple Scalar Fields

Category: E_2^{mS}

It is possible to display multiple scalar fields using techniques similar to those developed for high-dimensional scatter plots. Bergeron and Grinstein [Bergeron89] display multiple scalar fields over a region by an extension of the image display technique. Rather than just code the cells of the image by colour, they use an icon at each cell. This icon has attributes of colour, shape and even sound, and this has been used successfully to display five scalar fields simultaneously.

Scalar Field using Lighting and Shading

Category: E_2^S

This relates to the situation of a scalar field defined over a geometric surface, such as the temperature on the surface of an aircraft wing. In general, any property represented as a scalar may be visualized. The geometric surface can be presented as either a collection of polygons or as a parametric surface. Some base graphics systems such as PHIGS PLUS - [ISO/IEC(9592Am1)91] and [ISO/IEC(9592-4)91] - include these among their primitives: Set of Fill Area Sets With Data, and Non-Uniform Rational B-Spline Surface with Data.

The addition of light sources and thence smooth shading across the facets (Gouraud or Phong) can enhance the user's perception of the surface, but visual ambiguities can arise from the dual use of shading levels; see "Perception of 3D" on page 75.

Chang [Chang89] gives an interesting review of methods in this area, including references on interpolation over geometric surfaces.

Note: This is categorised as type E_2^S since a surface patch can be defined in 2D parametric space.

Bounded Region Plot

Category: $E^S_{[2]}$

There are several applications where the entity to be displayed is defined over a set of regions. For example, the density of population for different countries on a map. This type of display is known as a chloropleth map, or a bounded region plot. The value associated with the region is coded by means of colour or shading. (Indeed one can almost consider a shaded contour map as being of this type.)

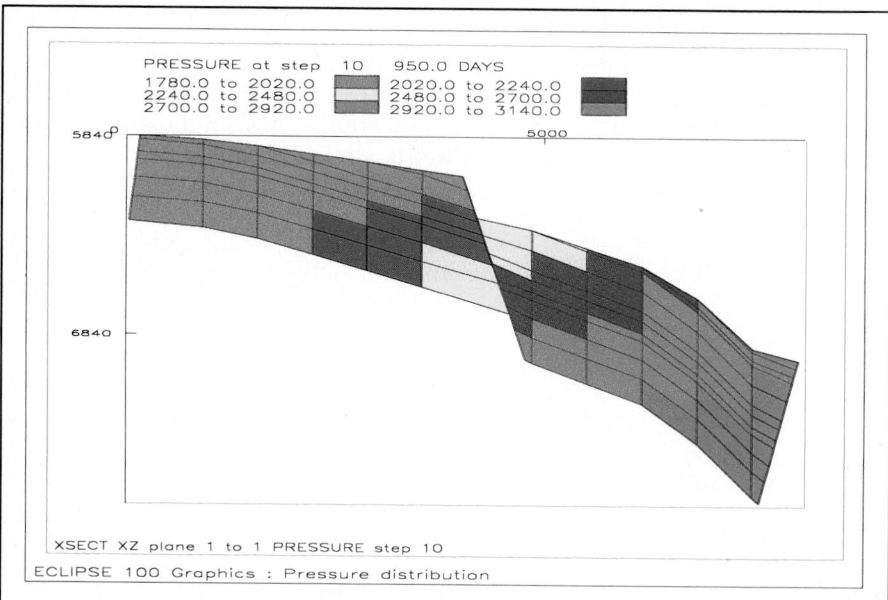

Figure 21. Example of a Bounded Region Display: The picture shows the value of pressure for each grid block in a vertical cross-section. The pressure is considered uniform over the whole grid block by the simulator, hence the need for this type of display. *Produced by INTERA ECL Petroleum Technologies using the Eclipse suite of oil reservoir simulation software*

Figure 21 shows an example of a single bounded region plot and Figure 22 on page 57 a more complicated example with two regions.

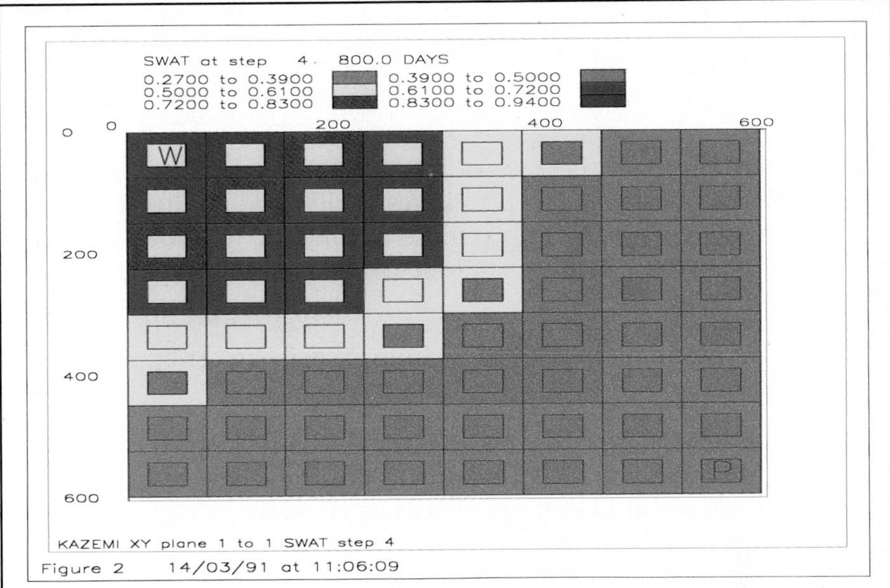

Figure 22. Example of a Multiple Bounded Region Display: The picture shows oil saturation for each grid block in a vertical cross-section. The rock is a mixture of two components with different porosities (like fractured limestone) so the oil moves at a different rate through the two components. *Produced by INTERA ECL Petroleum Technologies using the Eclipse suite of oil reservoir simulation software.*

3.5.3 *Three-Dimensional Domain*

This is the case where a function $F(x_1, x_2, x_3)$ is sampled at a number of points. In many applications, the data is supplied at each point of a regular lattice in 3D, and the general area is termed volume visualization.

Three techniques are described: two provide isosurfaces, that is they display surfaces of constant value within a volume; the other, called direct volume rendering, aims to display the characteristics of the volume itself.

Isosurfaces

Category: E_3^S

The general approach is to apply a surface detector to the sample array, and then fit geometric primitives to detected surfaces. Surfaces are then rendered by any appropriate technique. Methods for fitting geometric primitives include contour connection methods, and voxel intersection methods.

Contour connection methods include those by [Keppel75] and [Fuchs77].

Voxel intersection methods have been developed by [Lorensen87], [Cline88], and others. These latter algorithms are referred to as Marching Cubes and Dividing Cubes respectively.

Marching Cubes works by classifying the eight node values of a voxel as either greater-than or less-than a threshold value. The eight bits are used as an index into an edge intersection table. Triangles are fitted to intersected edges and then output along with the surface normals at the triangle vertices.

Dividing Cubes works similarly but outputs "surface points" (with an associated surface normal) and so does not have to calculate intersections nor fit triangles. Another technique is known as the cuberille method [Herman79]. This first sets a threshold marking the transition between two materials and then creates a binary volume indicating where a particular material is present. Each solid voxel is then treated as a cube and its faces output as polygons.

Surface-fitting is usually slow, but rendering the resulting surfaces is fast, especially if rendering hardware is available. See Figure 23 on page 59 for an example from the study of Altzeimer's Disease.

Numerous optimizations and enhancements have been developed for voxel intersection methods. Contour connection methods are still used but less favoured. See, however, the basket weave technique described below.

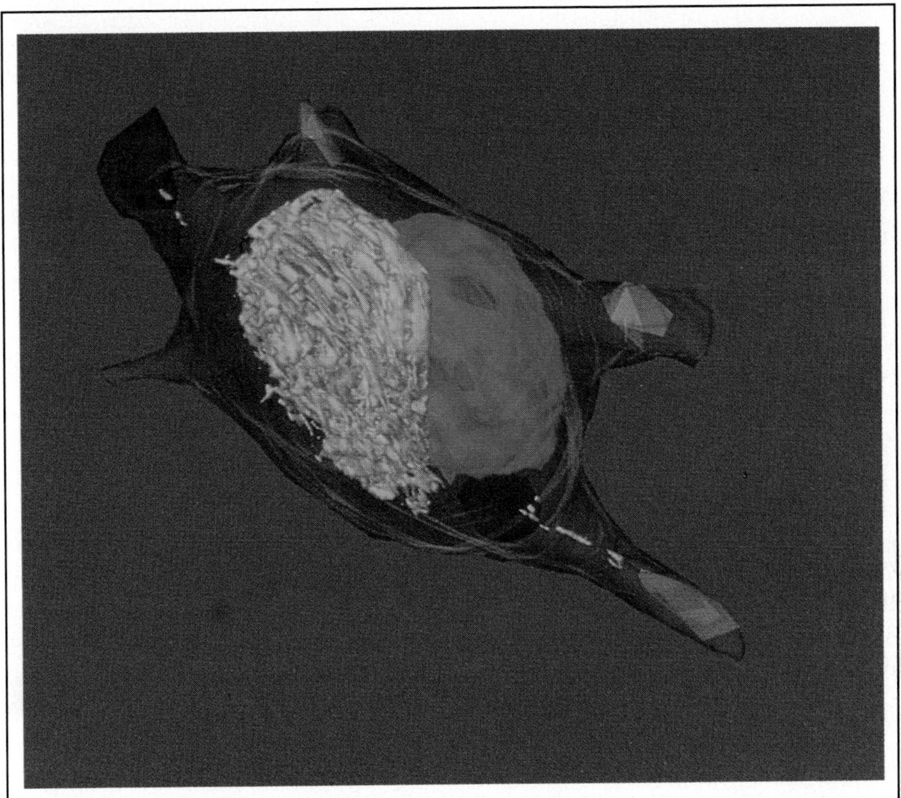

Figure 23. Example of Isosurface: This example is from the Study of Altzeimer's Disease. The picture shows Paired Helical Filaments (PHF) (orange) cracking cell nucleus (blue). *Produced by SYNU from San Diego Supercomputer Centre; data by Mark Ellisman (Univ of California at San Diego); visualization by Dave Hessler (San Diego Supercomputer Centre).*

Basket Weave

Category: E_3^S

Sewell [Sewell88] recently presented a new technique for the graphical representation of contour surfaces of a function of three variables. This is an extension of the isosurface technique to get around the problem of transparency. The algorithm assumes data is provided at points on a regular three-dimensional grid, with lines $x = x(i)$, $y = y(j)$ and $z = z(k)$. The algorithm involves drawing and projecting the contour curves corresponding to the two cross-sections $x = x(i)$ and

$y = y(j)$. The contours are drawn as thick opaque bands with hidden surface removal to give depth realism.

It is possible with this technique to display more than one level surface in a plot.

Additional scalar information can be colour mapped on to the bands, and so an entity of type E_3^{2S} can be visualized.

Volume Rendering - Ray casting

Category: E_3^S

Colour and partial opacity are assigned to each possible voxel value via classification tables (usually accessed with transfer functions). Images are formed by blending together voxels projecting to the same pixel on the image plane. Voxels may be projected in either object order or image order. Resampling along a viewing ray often requires finding a value between voxel nodes. Either tri-linear interpolation, tri-cubic interpolation, or nearest-neighbour are usually used in this case. Gradients are used to approximate surface normals and are usually found using central differencing.

An example using medical data is given in Figure 24 on page 61.

3.6 Techniques for Vector Entities

Very often, in applications which involve flow, the data consists of vector values at the geometric data points. The vector fields to be visualized may have two, three or more dimensions and exist in planar or volume fields.

The difficulties of vector field visualization are well expressed in [Helman90]:

> *We cannot directly display vectorial data on a two-dimensional screen, for example as a set of little arrows, and still interpret the result with the same ease as we would a scalar image. Our visual systems simply are not well adapted to interpret large volumes of vectors in this way, whereas we have superb abilities for understanding and interpreting images or depth-cued surface displays.*

One option is to extract some scalar quantity from the vector field, and use some established technique for displaying scalar data, such as described in the previous section. But there is typically a serious loss of information when this is done. There is a need therefore to tackle the vector visualization problem directly.

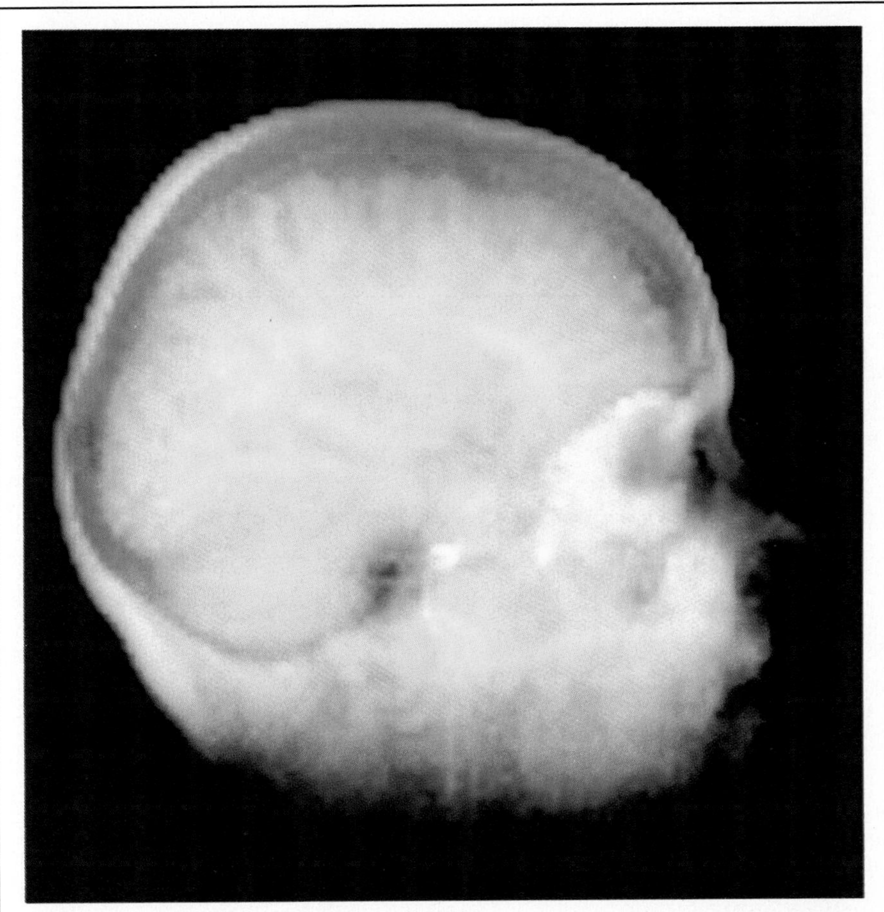

Figure 24. Example of Volume Rendering: The picture shows a volume rendering of medical data of a human head. The data set consists of 109 slices of 256x256. Each slice was reduced to 128x128 for visualisation. *Data from 'Head data' from the Chapel Hill Volume Rendering Test Dataset Volume I, obtained from the University of North Carolina, originally from Siemens Medical Systems Inc, Iselin, New Jersey, USA; visualization by Janet Haswell at the Rutherford Appleton Laboratory using the AVS system.*

3.6.1 *Two-Dimensional Vector Fields*

For 2D vector fields, some success has been achieved with relatively simple methods.

2D Arrow Plots

Category: $E_2^{V_2}$

It is typical to display vector data by the use of arrows. In the simplest case, a 2D vector in the plane can be visualized by 2D arrows [Fuller80].

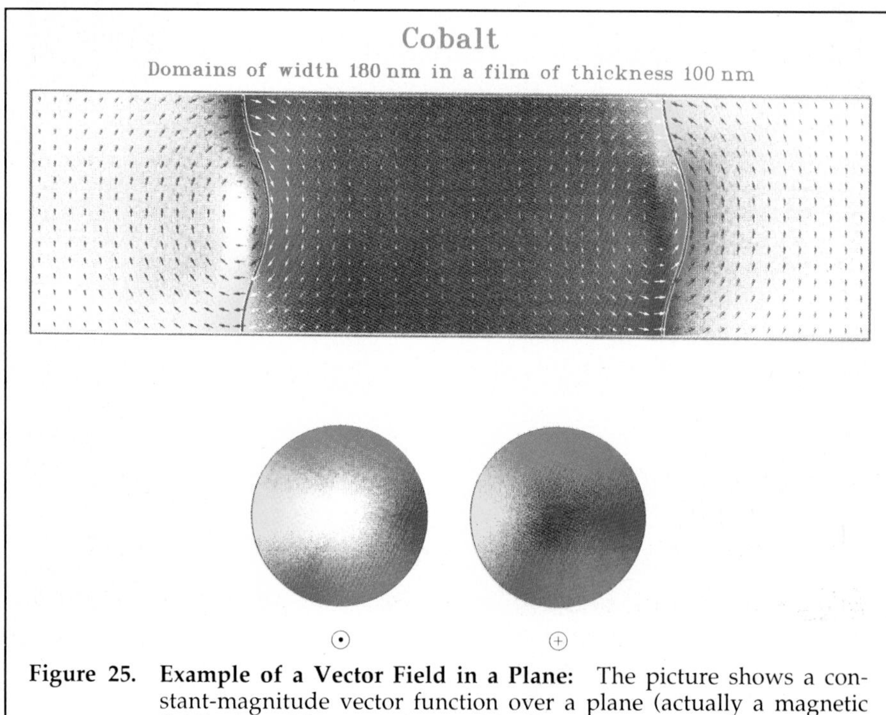

Figure 25. Example of a Vector Field in a Plane: The picture shows a constant-magnitude vector function over a plane (actually a magnetic field) plotted by using hue and lightness to represent the theta and phi components of the vector (only fully saturated colours are used). The colour transitions have been smoothed by introducing a random element into the colour choice. The arrows represent the same vector function, the apparent variations in length representing vectors into or out of the plane. *Picture by Malcolm Austen (Oxford University Computing Service).*

Figure 25 shows a vector function being represented in this way, superimposed on a field rendered using hue and lightness.

2D Streamlines and Particle Tracks

Category: $E_2^{V_2}$

An alternative representation is to show streamlines or particle tracks. These are static polylines which indicate the direction of flow in CFD, for example, or flux in electromagnetics. They often allow a greater insight into the data than a simple vector field.

The method used to construct the streamline may be a simple technique of choosing a starting point, and then constructing line segments by means of averaging the nearest vectors, or more sophisticated application-dependent techniques such as tensor-product spline fitting. Colour may also be added to indicate the velocity of the particle at each point in the streamline.

A difficulty with both arrows and streamlines is the density of information that is presented to the viewer. In the following method [Helman90], only significant streamlines are shown.

2D Vector Field Topology Plots

Category: $E_2^{V_2}$

The critical points of the field are identified by analysis of the Jacobian matrix of the vector with respect to position; saddle points, attracting nodes, repelling nodes, etc. are picked out. Streamlines are drawn from each appropriate critical point. The result is typically a very simple and uncrowded plot, but one from which an observer can infer the entire vector field [Helman90].

Of course time is very often a parameter in fluid flow visualization. The topology plots, being relatively uncrowded, can be extended to display change of topology with time: the streamlines are extended to ribbon surfaces.

3.6.2 *Three-Dimensional Vector Fields*

This is a hard problem. The first method is a simple but useful one for visualizing 3D flow in a planar cross-section of a volume.

3D Arrows in Plane

Category: $E_2^{V_3}$

A 3D vector in a 2D plane is displayed [Fuller80]. The arrows can point into or out of the display surface.

3D Arrows in Volume

Category: $E_3^{V_3}$

A 3D vector in a volume is displayed. Depth realism can be provided by depth-cueing or lighting/shading for solid vectors. Colour may also be mapped onto the vector to provide extra scale information, or representation of a separate variable. The choice of vector type and arrow head is discussed in [Kroos85]; it can affect the perceptibility of the direction of the vector.

3D Streamlines and Particle Tracks

Category: $E_3^{V_3}$

As with 2D fields, streamlines or particle tracks can be used. The visualization object may be a single polyline rendered in space, or a bundle of streamlines combined to form a ribbon. Visualization with ribbons often allows improved perceptibility of the way the particle tracks are integrated together. See [Helman90] and [Eaton87]. Key references are [Helman90] and [Richter90].

3.7 Techniques for Tensor Fields

Tensor fields arise in application areas such as CFD and Finite Element Stress Analysis. A second order tensor in 3D is represented by nine components, arranged in a 3x3 array. A tensor field would consist of a number of these arrays at each point in a 2D or 3D field. Mapping tensors to scalars is possible but loses much information. However direct visualization of tensors is not easy. Example techniques are the Lame stress ellipsoid and the Cauchy stress quadric [Fung65]. A new visualization technique is described by Haber and McNabb [Haber90], and summarised here.

Symmetric, Second Order Tensor Display

Category: $E_3^{T_{3:3}}$

From the symmetric 3x3 matrix, eigen analysis gives the tensor's principal directions and magnitudes. A cylindrical shaft is oriented along the major principal direction; the colour and length of the shaft indicate the sign and magnitude of the stress in this direction. An ellipse wraps round the central portion of the shaft and its axes correspond to the middle and minor principle directions of the stress tensor. The colour distribution of the disk indicates the stress magnitude in each direction.

For an illustration of the display of a tensor field, see figure 8 in Haber and McNabb's paper in [Haber90].

3.8 Image Processing Techniques

3.8.1 Introduction

Image Processing is a term given to a particular class of operations applied to regular two-dimensional rectangular arrays of data. Each array data value is commonly called a pixel or picture element. Historically, films were scanned to produce small blocks of measured intensity, resulting in a series of numbers which, if displayed on a screen, reproduced the original image. These images were usually rectangular, with a limited range of intensities for each point. A large number of algorithms have been developed for processing such data, and may be usefully applied to any two-dimensional data array as suggested by Pickover [Pickover88]. These algorithms can be found in subroutine libraries, or in a package framework; increasingly, implemented in hardware (particularly parallel processing hardware).

Applications include astronomical and terrestrial remote sensing, medicine, industrial inspection, etc., utilising many different algorithms. In this document, we are concerned only with those algorithms concerned with Scientific Visualization, and, as such, may cover three major areas: image enhancement, feature extraction and transformations.

3.8.2 Image Enhancement

Image enhancement within a scientific visualization system is the emphasis of image features for analysis or display and includes many techniques, amongst them the following.

1. Pixel Operations

   These are operations on individual data elements without reference to any other surrounding pixels. Such operations may include

   - Contrast manipulation

     This consists of a linear transformation on a range of field values delimited by user-controlled maximum and minimum output values. It is used to increase image contrast or to invert an image. The image can also be ***clipped*** of noise if the input signal lies within a known range, can be ***thresholded*** to produce a binary output, and ***sliced*** to segment image features contained in different grey-level regions.

   - Histogram manipulation

     The first operation is to construct the histogram of all the grey-level intensities within the two-dimensional array. The histogram is a representation of the distribution in the image. Based on this and the desired distribution (for equalisation, this would be uniform), a lookup table is produced. The pixels are modified to produce the desired distribution of grey-levels between the user-controlled minimum and maximum, typically with the changing image being simultaneously displayed. This technique can be useful for enhancing low-contrast images whose grey-levels lie within a narrow band (for example, Figure 26 on page 67, part (c)).

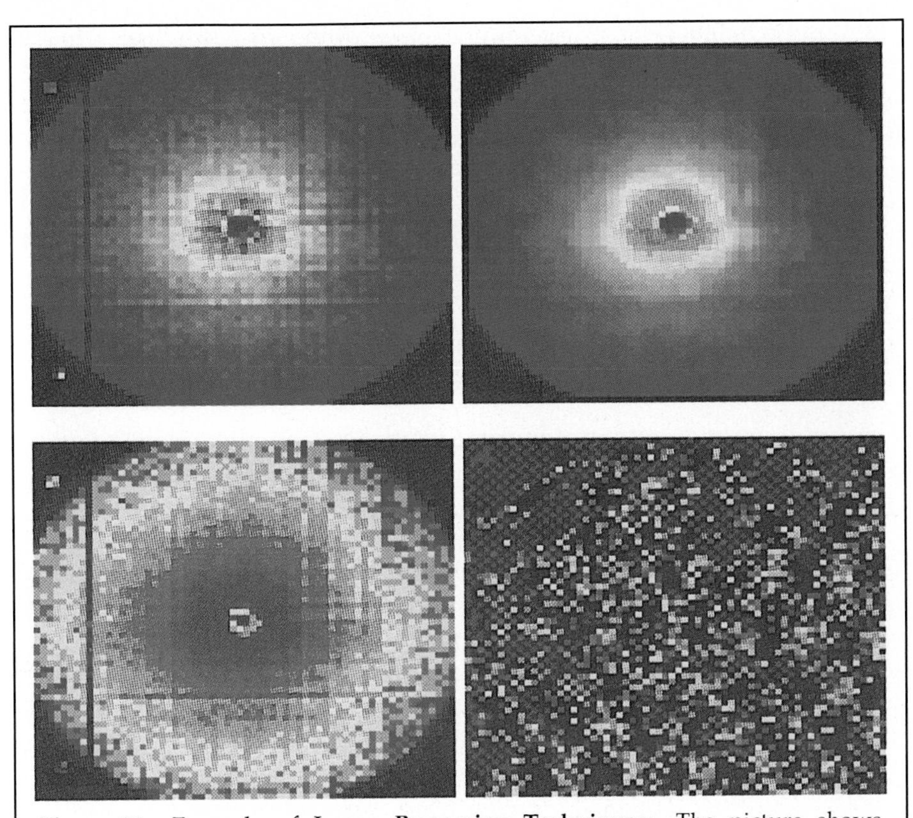

Figure 26. Example of Image Processing Techniques: The picture shows two-dimensional data taken from a liquid polymer exposed to the neutron beam in the small angle neutron scattering instrument LOQ of the ISIS Facility at the Rutherford Appleton Laboratory. (a) shows the original data where counts from the neutron detectors are displayed as a grey scale (or false colour) image, the hundreds of counts have been mapped into the colour scale shown on the left. (b) shows the result of smoothing the data with a median filter. (c) shows contract enhancement by histogram stretching while (d) is the result of applying a Fourier transform to the data. Although it looks completely different, it carries the same information as that in the other parts of the figure, and this image may be easier to measure or compress.

2. Local Area Operators

These are operations on the local area surrounding the pixel, often implemented as a sliding window moved across the data set. This window may be thought of as a filter centred on an image pixel which is replaced by a pixel whose characteristics are derived from the nature of the filter. Various filters include low-pass filters (useful for smoothing edges), high-pass filters (useful for emphasising edges), and smoothing filters (e.g. median filter as shown in operation in Figure 26, part (b).

3. Pseudocolour (False Colour)

   Pseudocolour can be thought of as mapping a set of grey-scale images into a colour image so that discrete features of the data set can be distinguished by different (non-natural) colours.

3.8.3 *Feature Extraction and Segmentation*

Feature Extraction allows quantitative measurements to be made from an image in order to describe it numerically. Segmentation techniques are used to isolate the derived object within the image.

Feature extraction makes use of several techniques based on underlying ideas such as spatial features (e.g. an object's features being represented by grey-level amplitudes), edge detection (where an edge is detected when adjacent pixels have grey-level discontinuities), boundary-extraction and contour-following (where boundaries are connected edges and can be found by tracking pixels that are connected), shape features (e.g. an object's features being represented by geometrical constructs), and texture (which may be thought of as structured surface patterns on an object).

Segmentation makes use of several techniques such as thresholding, labeling connected pixel sets, boundary-based profiles, and template-matching.

3.8.4 *Transformations*

Transformations involve the mapping of input data via some function, in order to compress it or more easily determine the features of the original data, e.g. detection of circles by the Hough transform. The result of this transformation may look very different from the original image, but measurements on the transformed image may give information about features in the original image. For example in Figure 26 on page 67, a Fourier transformation has been applied to the original image (a) creating that shown (d).

3.8.5 *Image Processing Techniques*

Some scientific visualization systems incorporate some of these image processing techniques, while in others, it may be necessary to use image processing libraries from other sources (e.g. IPAL [Crennell89] and [Maybury90], Spider [Takamura83]).

3.9 Animation

3.9.1 *Animation Techniques*

Animation is a useful tool for enhancing many types of visualization. Most viewing, lighting, data-classification, colour, and even data parameters can be varied between frames and using animation techniques often provides greater insight than using static images for data analysis.

Animation is a method for simulating continuous phenomena by displaying a discrete collection of images. By rapidly updating the display, the viewer is given the impression of watching a continuous event. In traditional animation, a large number of images, or cels, are hand-drawn and then filmed one frame at a time on a special movie camera. With the advent of computer animation, animators are only required to draw keyframes. The computer is then instructed to interpolate to find the frames between the keyframes. In scientific visualization, a similar process is often used, whereby the user specifies the situation for key frames in the sequence to be generated and the computer interpolates between them.

Using animation techniques is often crucial for understanding complex data, scenes, or phenomena. Projecting a scene of three or more dimensions onto a two-dimensional display requires collapsing objects of more than two dimensions into two-dimensional objects. If done with care, this projection can be performed so that the resulting two-dimensional image provides the viewer with three-dimensional cues. Clues to what has been "lost" by viewing in 2D may be provided by techniques such as depth-cueing, shading, perspective, and hidden surfaces. When three-dimensional objects in a scene are moved in relation to each other, or in relation to other fixed scene attributes, these cues are greatly enhanced.

Ideally any scene could be animated in real-time under user control; however, creating full screen images at adequate speed is not feasible with most currently available computer systems. Small scenes composed of simple geometric primitives can be animated in real-time on powerful graphics workstations, but screen update rates drop drastically as soon as scenes become even slightly complex. Animation under user control additionally allows scientists to explore a dataset via a real-time path of interest, rather than prespecifying a path to follow while rendering a batch animation.

Visualization of two- and three-dimensional scientific data usually requires the creation of images one at a time and then either playing them back on the computer using a flip book program, or else recording them to a medium, such as videotape, that allows fast playback of large numbers of images. The images themselves may have been created by any of the methods described elsewhere in this chapter.

Most animations show a temporal sequence, but animations can be controlled by any parameter when dealing with scientific data. By varying a parameter incrementally and creating an image for each new value of the parameter, it is possible to view how the other data are affected by the changing parameter. An example is varying temperature and observing how pressure on a three-dimensional grid changes, or varying time and observing how two fluids mix during a computational fluid dynamics simulation. When dealing with empirically-generated data the scientist has varied some physical parameter in order to create an interesting event for measurement.

From computer graphics literature we know that real-time motion is important in the comprehension of complex three-dimensional scenes. Real-time motion is particularly important for scientific data, where features lack the intuitive relationships inherent in geometrically defined object.

3.9.2 *Video*

Video provides a very cost-effective way of storing animation sequences. It provides a wide colour range and is easily viewed by one or many people simultaneously.

Video may be stored on videotapes or videodisks. By far the larger number of playback facilities will only support videotapes and so these are most commonly used for playback. Video disks (either magnetic or optical) are common in the larger video production facilities.

The cheapness of the video medium means that, if the video system provides the facility, it can be useful to record sequences at different speeds, to simplify the search for time-dependent phenomena of widely different frequencies in the same system.

Since television is a domestic commodity, video techniques are designed to meet the requirements of this market and there are aspects of video that are not perfect for scientific visualization. The most obvious is that video is designed for animated, low resolution pictures and not for static, high resolution ones. As a result, the viewer automatically expects pictures on a television screen to be animated.

A more surprising aspect is that it is very much harder to concentrate on a silent video sequence than on one with some soundtrack; whether this is an appropriate commentary or a not inappropriate music track is of less importance than the absence of silence.

It is tempting for newcomers to the use of video to start by producing an "animated" sequence designed on their regular graphics device, which will often be a medium or high resolution screen. This will usually produce a most disappointing result, for a number of reasons:

- many displays from visualization are extensively annotated; in a video, if the annotation is constant it is probably unnecessary (being better displayed on a leading title frame) and if it is changing with every frame it will probably become an unreadable blur on the video;

- most displays make some use of primary (red, green, blue) and complementary (cyan, magenta, yellow) colours; all of these at full saturation, are outside the colour gamut that can be encoded in any standard (PAL, NTSC or SECAM);

- most displays will use fine lines for delicate detail (such as the wiremesh superimposed on a contour plot; these are too fine to survive the bandwidth limitations of current video recording (or broadcast transmission);

- virtually all displays will have been designed for careful examination, with multiple points of interest; on video, more than one focus point is nearly always a disaster.

This highlights the conclusion that video is its own medium, superb for the presentation of animated results, using the colour gamut that is appropriate (PAL and NTSC have different limits in this area), but that (as any film-maker will know) the best shots do not just happen: they are planned and exploit the medium.

3.10 Interaction Techniques

It is arguable that the main feature scientific visualization systems provide, that "traditional" graphical display systems generally do not, is the ability to interact with the data. This interaction may involve only the post-processing of data or may be used to steer the generation of new data. Some of the techniques used may result in the modification of:

- general viewing,
- display technique,
- parameters,
- associated data,

but could also be used to select areas of interest and to annotate images.

3.10.1 General Viewing

This allows the user to change generic properties for a view. For 2D images this includes panning and zooming (in and out). Representations of 3D data, including volume renderings, could have additional control over lighting (colour and position) view point/camera position, and, for surface representations, shading technique used (wireframe, Gouraud or Phong). Animation sequences could be generated for 2D images by defining a sequence of panning and zooming operations and for 3D a sequence of 3D transformations (both on the object and on any light sources used); these may be defined one frame at a time or by recording an interactive session. If the viewing position is an integral part of the input data it may be a requirement to create a new data file with the chosen viewpoint (in 2D or 3D).

3.10.2 Display Techniques

This covers the changing of the visual representation of the data until the most suitable representation is found, e.g. generating line contours for a 2D image then switching to a mesh representation (where the values used to generate the contours are converted to height values in a 3D mesh). This may involve more than one view showing the two representations, or the two techniques could be combined into one view (it may be possible to enhance the visual information by overlaying the original coloured image).

3.10.3 Parameters

A number of applications can benefit from user control of parameters not only for the display but also where the data generation can be controlled by parameters. It might be desirable to change the style in which vectors are displayed (as arrows, cones, tubes etc.), or perhaps their density, until the required information is suitably displayed. Similarly increasing the number (possibly by interpolation) could allow increased detail when homing in on areas of interest.

For particle tracing, the user may specify the number, speed and pattern (e.g. grid) of the trace, or the object used in the trace (e.g. spheres, arrow, ribbons).

In volume visualization, by changing opacity levels for different data values it is possible to make outer layers of a volume transparent and make features within the volume apparent. Interactively changing these levels allows the user to see the required combination of outer and inner details.

The visualization of equations may require parametric input before the data can be generated, and parameters could be used in deciding mesh size before Finite Element Analysis.

3.10.4 Associated Data (such as Colourmaps)

Interactive control of colourmaps can aid considerably in the exploration of image based data (including volumetric data). Either generating one's own colourmap (e.g. using AVS colourmap manager) or changing the range of data to be mapped onto colourmaps (clamping, histogram stretch) can make visible hidden artefacts. Allowing the user to choose a colourmap can help overcome differences in perception of colour. Colour-blind individuals could assimilate more information without error from an image created using grey-scale mappings than from a multi-coloured image.

3.10.5 Selection

It is possible to select 3D volumes, 2D areas, lines or points from the graphical representation in order to reduce the complexity of data displayed. One technique is to display the simplified image, another is to display the data value(s). The term data probe is generally used to indicate the latter. Data probes in 2D space use a cursor which the user positions over the image at a point of interest. This location can either return the intensity of the pixel which could be mapped to the data value at that point, or more likely be used to map back to the x,y coordinates which can in turn either map back to the original data values or possibly to interpolated values.

3D data probes exist but perhaps a more systematic approach is to reduce the 3D data set to 2D then use the simpler 2D data probe. By interactively moving a cutting plane through a volume the required 2D slice (which would not necessarily have to lie in x, y or z planes) could then be displayed in a separate window for further processing. This approach could cause difficulties in getting back to original data values (if interpolation is used to generate intermediate values in image).

Visual3 (MIT code primarily for the display of CFD data) uses this approach and takes it to an additional level by allowing the user to select a line through the image for the creation of a graph (data value vs. length of line).

The NCSA tools allow the user to reflect the cursor position back to a table of data values (adjacent values are also visible), and also to move around the table of data values and have the image updated accordingly.

3.10.6 Annotation

Having generated the required view(s) for a data-set the user will often need a recorded version, as a CGM or PostScript file, video or hard copy. For reference purposes additional information, such as labels, titles and the addition of legends can make the recording more complete - these can be interactively positioned by the user.

3.11 Perception of 3D

As well as considering the visualization techniques themselves, it is essential to consider the user's understanding of the application when these techniques are used. Chapter 5 (Human-Computer Interface) considers the interface to visualization systems in a general context. Here we direct ourselves to a specific issue: when the domain of the visualization object is three-dimensional or greater, how does a user visualize what is intended when the information is presented on a two-dimensional screen? Perception is the process of reading non-sensed characteristics of objects from available sensory data. It is difficult to hold that our perceptual beliefs - our basic knowledge - are free from contamination. We not only believe what we see: to some extent we see what we believe.

The basic question is: What kind of 3D object does this 2D surface show? The problem is acute as any 2D image could represent an infinity of possible 3D shapes. Extra information is essential, and a number of possibilities exist.

Context
: Additional information surrounding the object can help the recognition process.

Perspective, lighting and shading
: Objects and their shadows appear smaller the further away they are.

Stereo Views
: The merging of two views by the brain gives important depth-cues.

Movement
: The addition of motion can help in visualizing the extra dimension.

3.11.1 Context

Looking at picture (a) in Figure 27 on page 76 we see an ellipse; however, the identical object, seen in the context of picture (b), is clearly a circular ring! The extra information of the young boy holding the hoop is essentially independent of the information that is being conveyed.

Figure 27. (a) An ellipse; (b) a boy with a hoop: *Picture from Terry Hewitt, Manchester University.*

As a further example, look at Figure 28 on page 77. These are from the JADE (JApan Deutschland England) experiment studying electron-positron interactions at DESY, Hamburg, Germany. To give these figures some scale, picture (b) represents the detector area, which is the size of a small house and weighs 1200 tonnes. Each of the brown boxes represents an iron-loaded concrete block and typical dimensions are 5 metres cross-section. Picture (a) conveys little information, but in pictures (b), and also (a) and (b) in Figure 29 on page 78, additional information is added to help the experimenter understand the picture.

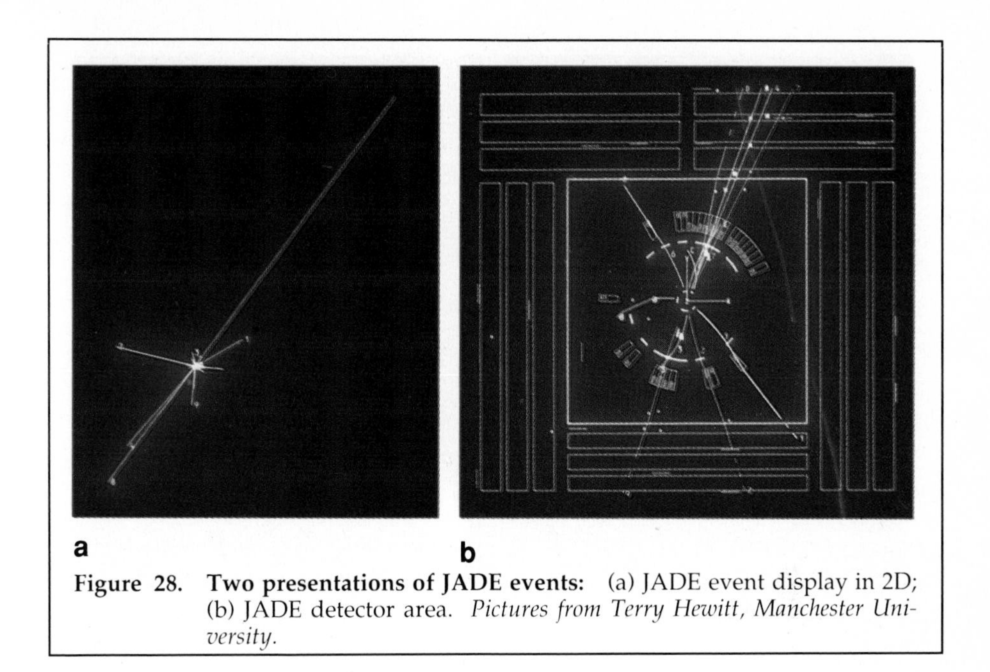

a b

Figure 28. Two presentations of JADE events: (a) JADE event display in 2D; (b) JADE detector area. *Pictures from Terry Hewitt, Manchester University.*

The concrete blocks in picture (b) of Figure 28 are there actually to protect the experimenters from stray radiation and nothing to do with the information collected; they are in the picture to help the physicist orientate himself with relation to the physical layout of the equipment and hence the physics.

3.11.2 *Perspective and Lighting and Shading*

It is a basic fact that objects further away appear smaller. Once again pictures from the JADE experiment illustrate this. Picture (a) in Figure 29 on page 78 includes a box without perspective, and it does not look quite correct. It is possible to see the Necker effect. The picture can be seen in two ways: one looking up at the bottom of the cube, and one looking down on to the top of the cube. This ambiguity can be removed using a perspective projection as shown in picture (b).

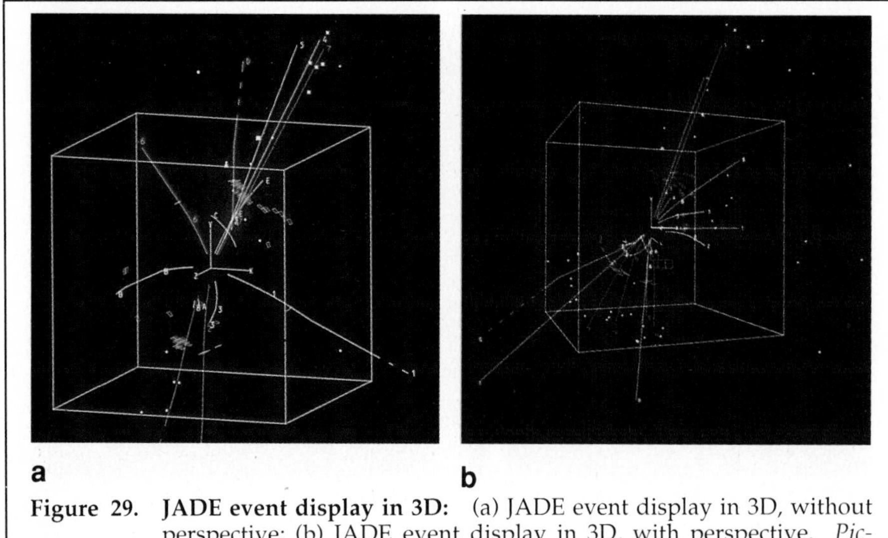

Figure 29. JADE event display in 3D: (a) JADE event display in 3D, without perspective; (b) JADE event display in 3D, with perspective. *Pictures from Terry Hewitt, Manchester University.*

An interesting aside is the Ames Room: the further wall is sloped back at one side so that it does not lie normal to the observer; perspective is used to give the same retinal image as a normal rectangular image. We are so used to rectangular rooms that we accept it as axiomatic that it is the objects inside which are odd sizes. The brain makes the wrong deduction. The odds are easily changed. It has been reported that a newly married wife will not see her husband shrink as he walks across the room, but instead will see the room approximately its true, peculiar shape! - see [Gregory70].

There is more to perspective than just geometry. Increasing haze and blueness with increasing distance are also important, and these effects are realized by depth-cueing in systems such as PHIGS PLUS, and examples without and with depth-cueing are shown in Figure 30 on page 79 and Figure 31 on page 79.

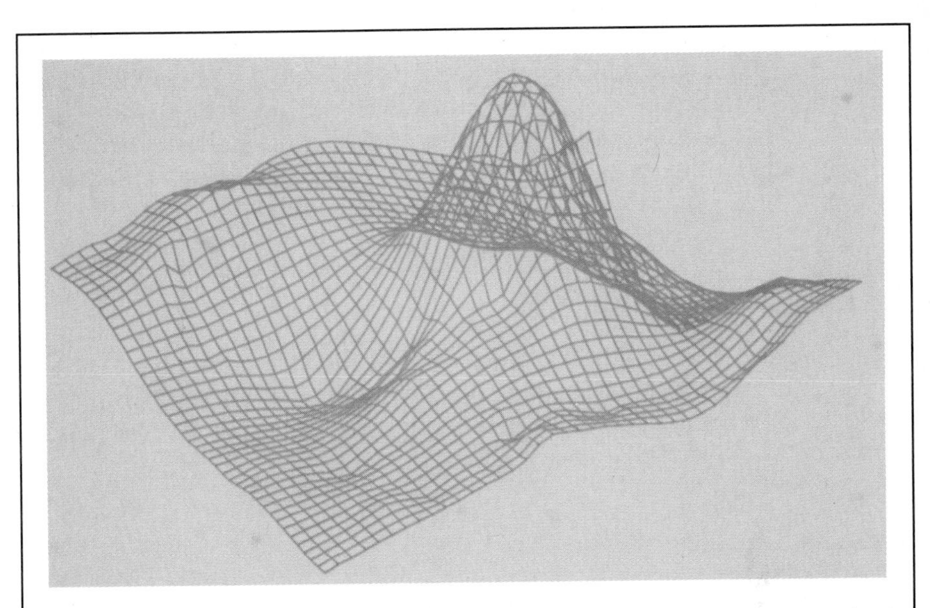

Figure 30. Wire-mesh surface, drawn without depth-cueing: *Picture from Terry Hewitt, Manchester University.*

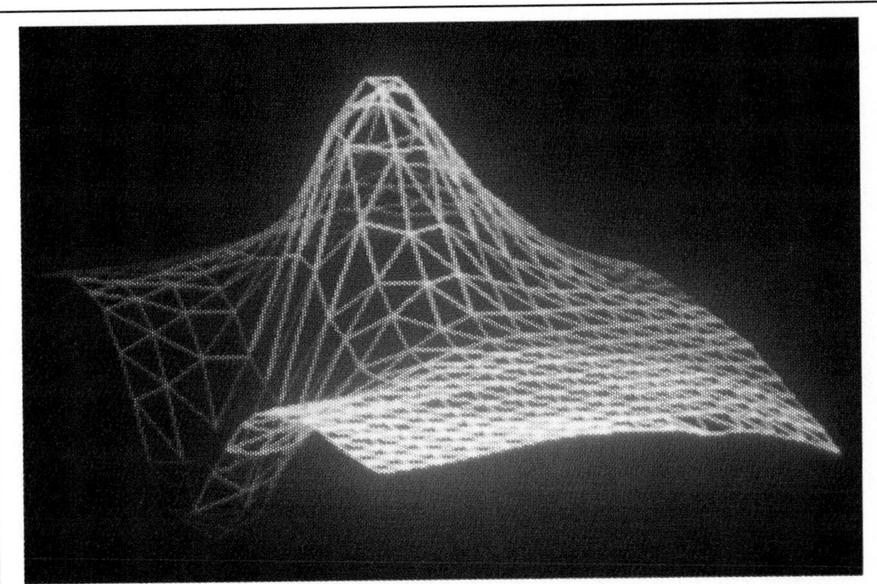

Figure 31. Wire-mesh surface, drawn using depth-cueing: *Picture from Terry Hewitt, Manchester University.*

Of major importance is the addition of light sources to the environment. Shadows are another cue, and though PHIGS PLUS supports a number of light sources, it has no explicit support for shadow generation. The resultant shading represents the orientation of objects, though ambiguities arise if colour is used to encode other information on the surface (for example, in a height field plot).

3.11.3 *Stereo Views*

When we are not looking at real objects the brain merges the two slightly different views from each eye to give one image in the brain. This stereo effect is a most important cue to depth. It can be simulated on displays in several ways:

Red/Green images
A pair of glasses is worn, with one red lens and one green lens; a red and green image are superimposed on the screen. The left eye sees the red image and the right eye the green image, and the brain creates a stereo image. This is satisfactory for monochrome pictures, though can cause problems for those with colour defective vision.

Separate images
Figure 32 on page 81 shows shows two images side by side. Using either special glasses (or placing one's hand in front of one's nose), each eye should be focussed on the corresponding image. The two images should merge into a single stereo image.

Polaroid glasses
The above two techniques are adequate but unsophisticated. In the latest technology, a pair of glasses, with an infra-red detector, is worn. The lens can be made transparent or opaque. The screen displays alternate images, and the infra red detector arranges that the correct lens is clear.

Movement
Another significant cue to perceiving 3D objects is movement. In Figure 30 on page 79 the mess of lines makes it difficult to see what the object is. Rotating this continuously on the screen makes it easier to "see" the shape of this surface. Sadly it is not a symmetric operation: if instead of moving the object, the observer is moved (try shaking your head from side to side) the depth is not observed. A real example is a weather vane.

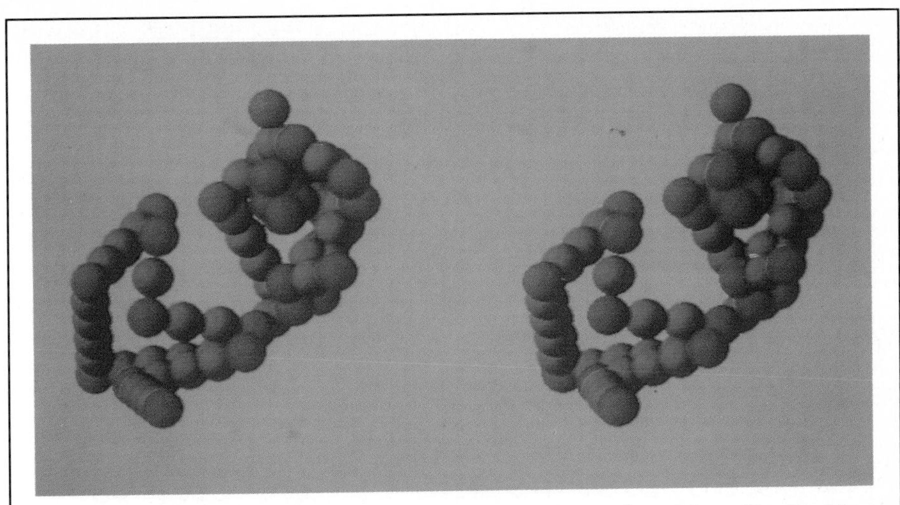

Figure 32. Twin views for stereo viewing: *Picture from Terry Hewitt, Manchester University.*

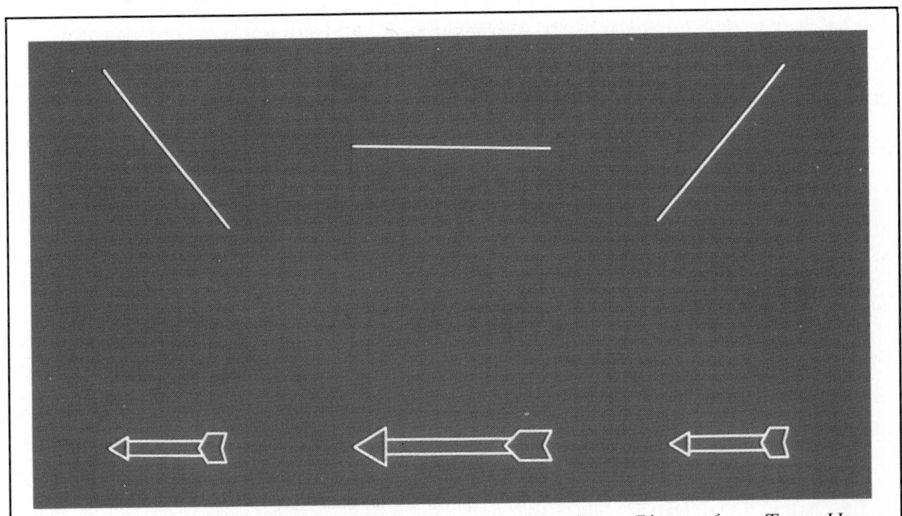

Figure 33. Arrows demonstrating visual ambiguity: *Picture from Terry Hewitt, Manchester University.*

Figure 33 shows in the bottom half, what might be seen from a distance. Three arrows pointing to the left, suggesting that the wind is blowing from the right. The plan view above shows that the left arrow is pointing towards the viewer, the centre arrow to the left, and the right arrow away from the viewer. The projection of the vane gives ambiguity as to its length and direction. As the vane rotates, its true length and orientation become apparent. As was noted earlier in "Techniques for Point Data" on page 43, movement has been judged the best way of conveying depth information in 3D scatter plots.

3.12 Conclusions and Future Trends

The main theme of this chapter has been to develop a structure to the wide range of techniques that are used in scientific visualization. The structure is based on the entity being visualized, and is divided on one axis by the type of entity (point set, scalar, vector or tensor field) and on the other axis by the dimension of the domain over which the entity is defined. Many of the common presentation techniques have been described, and classified according to the structure proposed.

Each technique can typically be seen as three separate stages: the creation of an empirical model from some supplied data; the mapping of that model to some suitable abstract visualization object; and finally the rendering of that object.

Many of the techniques are long established; the difference now is that in visualization as a subject, the techniques form just one component of an integrated system. However there are some subtle changes. In the past, the construction of an empirical model has been a very important step: often there would be very limited data, and a good interpolation process was required to "fill in" for the missing data. Hence higher order interpolation involving piecewise cubic polynomials has been important.

Today however the data collection process has become more powerful. Satellites transmit data at fine resolution; medical scanners too provide high density data. Thus the problem is not so much one of filling in missing data, but of filtering or smoothing data which already exists.

3.13 Key References

Most references are provided in the bibliography at the end of the book. A few references which are considered key for this chapter are provided here.

General

[Bergeron89]
Bergeron R.G., Grinstein G.G., "A reference model for the visualization of multi-dimensional data", *Proc. Eurographics '89,* **(Elsevier Science Publishers B.V, 1989)**, *pp 393-399.*

Presents a reference model for the visualization of multi-dimensional data. The fundamental concepts of lattices are introduced and explored within the context of PHIGS. The lattice scheme is based on classifying data rather than the underlying entity, and can be seen as complementary to the approach in this chapter.

[Helman90]
Helman J., Hesselink L., "Representation and Display of Vector Field Topology in Fluid Flow Data Sets", *Visualization in Scientific Computing*, ed. Nielson G.M., Shriver B., Rosenblum L.J., **(IEEE Computer Society Press, 1990)**, *pp 61-73.*

This gives an account of the difficulties of vector field visualization. It proposes the vector field topology method described above - in which aim is not to display ALL the data but just the significant parts.

[Nielson90]
"Visualization in Scientific Computing", ed. Nielson G.M., Shriver B., Rosenblum L.J., ISBN 0-8186-8979-X, **(IEEE Computer Society Press, 1990)**.

This volume is a valuable collection of papers on different aspects of visualization. Several papers are key references for advanced techniques - for example, Helman and Hesselink on vector field visualization, Crawford and Fall on multidimensional scatter plots, Haber and McNabb on techniques generally and tensor fields in particular, plus a host of other valuable papers.

[Richter90]

Richter R., Vos J.B., Bottaro A., Gavrilakis S, "Visualization of flow simulations", *Scientific Visualization and Graphics Simulation*, ed. Thalmann D., **(John Wiley, 1990)**, *pp 161-171.*

This gives a review of fluid flow visualization, with reference to three specific problems: hypersonic flows, flow in hydraulic turbines and turbulence.

[Thalmann90]

"Scientific Visualization and Graphics Simulation", ed. Thalmann D., ISBN 0-471-92742-2, **(John Wiley, 1990)**.

Another collection of papers on scientific visualization. Of note is the paper by Wilhelms on Volume Visualization, which is almost a guided tour around the references on the subject; and the paper by Richter et al on Flow Visualization with particular reference to CFD applications.

[Tufte83]

Tufte E.R., "The Visual Display of Quantitative Information", **(Graphics Press, 1983)**.

This delightful book explores the strengths and weaknesses of various forms of visual presentation, highlighting both successful and ineffective techniques and providing analysis of visual cheating.

Empirical Model Building

[Lancaster86]

Lancaster P., Salkauskas K, "Curve and Surface Fitting - An Introduction", **(Academic Press, 1986)**.

This volume gives a good introduction to interpolation, and the building of an empirical model from data. The emphasis is on interpolation, but there is some discussion of approximation when the data is subject to noise.

Volume Visualization

[CG90]

"Special Issue on San Diego Workshop on Volume Visualisation", *Computer Graphics* vol 24 (5), **(Nov 1990)**.

A collection of papers representing the latest work in this area.

[Levoy88]

Levoy M., "Display of Surfaces from Volume Data", *IEEE Computer Graphics and Applications* vol 8 (3), **(May 1988)**, *pp 29-37*.

This paper gives a very readable introduction to volume rendering, and its advantages compared with isosurface techniques.

[Levoy90b]

Levoy M., "A Taxonomy of Volume Visualization Algorithms", *Introduction to Volume Visualization: SIGGRAPH Course Notes*, **(Aug 1990)**, *pp 6-8*.

This provides a useful taxonomy of volume visualization algorithms.

Image processing

[Jain89]

Jain A.K., "Fundamentals of Digital Image Processing", **(Prentice Hall International, 1989)**.

Provides a general introduction to the techniques of Image Processing

[Pratt78]

Pratt W.K., "Digital Image Processing", **(John Wiley, 1978)**.

This is a good reference work for digital image processing; although now somewhat out-of-date, a revised version has been published in 1991, by the same author and publisher.

[Rosenfeld82]

Rosenfeld A., Kak A.C., "Digital picture processing (Second edition)", **(Academic Press, 1982)**.

The second edition of Rosenfeld's 1976 book on the same topic.

Chapter 4

DATA FACILITIES

Edited by Lesley Ann Carpenter

4.1 Introduction

> *"Today's data sources are such fire hoses of information that all we can do is gather and warehouse the numbers they generate."*

This quote taken directly from [McCormick87] well illustrates one of the major problems in developing effective visualization software and techniques; that of efficiently handling and processing vast amounts of data. Since 1987 the necessary technologies and hardware to analyse and present the information in ways which are useful to the scientist have moved forward but the need for an efficient data handling methodology remains at the crux of the visualization initiative.

An illustration of the extent of the problem can be taken from the discovery that the ozone layer protecting the earth has deteriorated to such an extent that there is a "hole" over Antarctica. Although this has only come to light in recent years, the evidence has actually existed in data (collected and archived) for over 10 years.

Effective solutions for handling vast quantities of data being input to a system, its storage and subsequent techniques for its processing and analysis are needed. There is no reason to suggest that the number of sources and amount of data will do anything other than increase, for example satellite resolution has increased ten fold in last few years and supercomputers formerly working at 0.1 to 1.0 gigaflops are now able to handle 1.0 to 10.0 gigaflops.

This chapter proposes a classification suitable for describing the different types of data flow identified within the reference model for visualization outlined in chapter 2 (Framework). Data transfer formats for the import/export of data are discussed, data compression techniques outlined and a brief overview of facilities for managing data is given. The chapter concludes with a number of recommendations for future work which it is felt is needed in this area.

4.2 Data Sources

A distinction can be made between data sources external to the computing system, of which the visualization system is a part, and those which are internally generated. External sources might include results generated by satellites (weather, military, astronomical, resource etc.), spacecraft, instruments (geophysical, oceanographic etc.) or scanners (medical, engineering etc.). Internal sources might include the retrieval of information from archive or the output generated by a numerical model (e.g. finite element analysis).

Ever improving data acquisition methods have led to an increased need to be able to deal with a continuous stream of data - "fire hoses". Visualization methods increasingly need to support the storage, processing and display of data in real time. This data is also being archived on a large scale - "warehouses".

Some of the features which need to be catered for include:

- Datasets consisting of multiple data fields
- Datasets of a very large size (up to one terabyte)
- Time varying data
- Datasets with disparate components and organization

4.3 Data Classification

Numerous classifications of data exist; these include methods based on data sources, formats, and types of data. For pragmatic purposes it was necessary to propose a data classification that could be used as part of the framework. The classification had to be broad enough to encompass both the data being investigated and the data required for the system to execute.

4.3.1 *External and Internal Data*

A distinction is made between external and internal data. It is acknowledged that any visualization system must be capable of importing data from external sources. There is also a requirement to export data from the system. The distinction between import and export data is somewhat artificial as the output of one system is likely to be used as the input to another. The important point to note is that the system must be able to communicate with the outside world and is unlikely to exist as a "totally enclosed" solution. Internal data, however, are that which is used wholly within the system. Data which control the flow of module execution, for example, is purely internal.

4.3.2 *Original and Derived Data*

The source of the data being analysed is important. A practical model to assist in classifying data sources needs to relate to the original data and any data subsequently derived from that base. For example, observations made of real world phenomena such as magnetic resonance imaging or remote sensing may contain a significant amount of extraneous "noise". As a result the data will often need to undergo some enhancement, or preparation, or "cleaning up" process prior to beginning formal analysis. This refined data is therefore ***derived*** from the original data. Further manipulation may occur, for example projecting from a higher to a lower dimension, and this too can be regarded as generating more derived data. It may be vital to provide an "audit trail" to track the methods by which the derived data was generated and also the location and availability of the original data in case there is a requirement for its referral or use in subsequent stages of analysis. Further examination and refinement may be required to enable maximum insight into the original data. The derived data should be labelled or tagged to identify it as non-original data. For example, a common approach to tagging derived image datasets is to include a history of all processing which it has undergone.

Original data is often referred to as raw data; use of this term has been deliberately avoided as it may carry implicit connotations about the quality of the data.

4.3.3 *Basic Primitive Elements and Logical Sets*

The most pervasive storage medium and manipulation method is digital, which itself imposes some restrictions on the data model. From the basic primitive element in a digital system, a bit, numerous higher level abstractions and structures can be constructed. These include, bytes, arrays, floating point, integer, and characters. The storage and manipulation of analogue information in a digital medium requires some form of analogue/digital conversion (in both directions) and a careful consideration of sampling and quantization.

The basic digital primitive can be employed in the construction of higher level abstractions which in turn can be used to build "sets" of logically grouped data elements. At this level structures such as arrays and mappings can be employed. Both basic primitive elements and logical sets provide abstractions for use by all other components of the system.

4.3.4 *Geometric and Property Data*

Geometric data
can be used to represent the shape of an object. This includes polygons, curved surfaces and simple coordinates. Shape can exist as an independent variable such as an aircraft wing to be coloured by temperature, or as a dependent variable as in the case of a finite element analysis of torsion on an actuator arm.

Property data
in its simplest sense, is non-geometric data. Properties are commonly items such as scalars, vectors, and tensors, for example representing perhaps temperature, velocity, or stress. The property is usually considered to be an dependent variable, and is commonly the actual area of interest or concern. Property data may be associated with geometry data.

Both geometric and property data range may be inherently structured in a number of different ways. For example, the lattice model of Bergeron and Grinstein [Bergeron89] allows for organization of a range of multidimensional data, although this model is not suited to all possible arrangements of data. One of the most obvious divisions to be made in data organization is between regular and scattered data. Regular data is present over a defined tessellation grid which may consist of cubes, tetrahedra, triangle, squares, prisms etc.. A number of other architectures exist, generally developed in response to specific needs. For example, Octrees and Quadtrees are becoming widely used, particularly in Geographic Information Systems - see [Samet88a] and [Samet88b] - and in those areas of image analysis concerned with the extraction of geometric shape (e.g. computer vision) [Rosenfeld84].

4.3.5 *Record Data*

A number of simple data types can be identified which may be stored in a simple record format. These include: metadata, command data, and control data.

Metadata

In order to adequately document the history of a dataset an audit trail may need to be kept which records any refinement, processing etc. that has been performed during the derivation of that data. Metadata can also contain information on:

- type and structure of the associated dataset
- attributes such as, field names, units, ranges, calibration, missing data, statistics, tolerance and accuracy information
- ancillary information such as, spatial location, time tags, audit trail(s)
- textual comments and records
- colour tables

Appropriate attention must be paid to the extensibility of metadata storage and the setting of default values. In the control and documentation of relationships metadata plays a crucial role, for example documenting the components of a set of geometries or properties over space and/or time.

Metadata may of course become, or have been, derived data.

Command Data

A fully interactive visualization system may, for example, be driven by a pointing device such as a mouse; this mode of operation being good for defining unique operational sequences such as those used in data exploration. When repetitive sequences are used it is useful to have the ability to record and play back these commands. The data used to drive the system in this way can be held as simple command scripts which may be full, verbose and text editable or abbreviated, terse and, possibly, encoded. In this way an audit trail of the actions performed by the user can be generated (and potentially archived) - thus information about the processing performed on the internal data can be recorded.

Control Data

contains parameter information necessary for the execution of modules within the system. By the nature of such information it can be in a variety of formats or standardized by the module implementor(s). Control data may be in clear text or encoded.

4.3.6 *Relationships*

Datasets can be logically grouped to build associated sets or formalize the dependence or independence between datasets. Some possible relationships include:

- transformations from the original data
- grouping of datasets with a common property, e.g. similar geometries
- linkage information

Linkage information is particularly important in disciplines such as molecular modelling where there is a need to be able to specify the bonding relationships between atoms. A link may also occur between datasets.

Relationship data must be held in a format appropriate to its specific nature, e.g. clear text for lists of linked files or floating point matrices for relative positioning. Some relationship data will necessarily be stored with the data whilst other data relations are likely to be stored as metadata.

4.4 Management of Data

At the 1990 SIGGRAPH meeting, Lloyd Treinish of IBM's T. J. Watson Research Center stated (when defining the terms of reference for a "Data Management for Visualization Working Group"):

> *The organization, structure and management of data to be analysed are critical for effective visualization*

Traditionally methods of handling scientific data have revolved around the use of flat sequential files, often with the obvious disadvantages of being inefficient for accessing subsets, access times and ease of use, particularly when large datasets are being considered.

Over the last 20 years or so the use of databases and database management systems (DBMS) has become widespread in commercial spheres but this trend has not been (generally) taken up by the scientific community. There are many reasons for this but these are largely related to the different needs of the two communities, for example many existing DBMSs:

- are built upon the relational model - this model cannot easily accommodate multidimensional data or hierarchical data structures. Neither do they generally provide adequate performance for the size, complexities and type of access required or predicted for scientific visualization.

- provide superfluous facilities - commercially orientated databases provide integral support for facilities such as elaborate update facilities, concurrency control, audit trails, report generation, transaction processing etc.. These type of facilities are sometimes required in the analysis, management and display of scientific data, but otherwise may impose unnecessary expense and resource implications.

- do not offer appropriate language support - traditionally, few commercial databases have provided support for scientific programming languages such as FORTRAN and C, Application Programmer Interfaces (APIs) being developed for traditional business languages such as COBOL; this trend is changing.

A number of database and data management systems specifically designed for dealing with scientific data have been developed in recent years, for example, Model 204 (Computer Corporation of America) and System 1032 (Access Technology Inc. (a CompuServe Company)). Likewise the traditional vendors of commercial databases are making renewed efforts to address the needs of the technical user, an example being INGRES. A number of institutions, especially in the USA such as the National Supercomputing centres and NASA, have developed generalised data formats/structures for the storage and exchange of scientific data. Examples are HDF (Hierarchical Data Format) developed at the National Center for Supercomputing Applications (NCSA), University of Illinois at Urbana-Champaign [NCSA89] and netCDF developed by Unidata, part of the University Corporation for Atmospheric Research, Boulder, Colorado - [Rew90a] and [Rew90b]. In Europe the RSYST system has been developed by the University of Stuttgart, Germany [Lang90].

Ideally a data management system, to be utilised efficiently by a scientific visualization system, needs to be able to provide support for some or all of the following:

- a common strategy for reading/writing data;
- organizational facilities for handling data;
- user "interaction with their data", i.e. "reasonable response times";
- a de-coupling of the management of data from the access of data;

- provision for controlling redundancy;
- the propagation of updates (to avoid inconsistency);
- archive facilities;
- security methodology;
- maintenance of the integrity of the data.

4.4.1 *Data Description and Manipulation Languages*

SQL is an International Standards Organization standard for a query language for Database Management software [ISO(9075)87]. It was developed by the IBM Corporation in the late 1970s as a standardized language interface for data definition and manipulation within a relational environment. Many commercial relational database companies have adopted (or are at least offering support for) SQL as the user interface to their systems. More recently, the role of SQL has been extended further and it is being used as a "bridging" language between applications built upon databases. ANSI and ISO have begun work on SQL2, a new version of the SQL language which will take into account increased demands of users and the support of client/server architectures. SQL statements are divided into three language sub-groups:

- Data Description Language (DDL)
- Data Manipulation Language (DML)
- Modules and Procedures

The terminology introduced by the SQL standard has become widely used and as such will be adopted as the basis for the structure of the remainder of this section.

4.4.1.1 *Data Description Languages*

There have been a number of attempts to define a generic syntax for describing data, these being known as Data Description Languages (DDLs). Such data description languages have to deal with:

- elemental data types, such as integers, character strings etc.,
- machine representation of the elemental types (which could, for example, include the ordering of bytes within words)
- the structure of sets of basic primitives

As well as DDL - a data description language for declaring the structures and integrity of an SQL database [ISO(9075)87], other languages have been developed for describing data for example the External Data Representation Standard (XDR) developed by Sun Microsystems [Sun88]. A number of application-specific languages have also been developed, for example in image processing [Duff81] and physics [Vandoni89].

4.4.1.2 *Data Manipulation Languages and Macro Facilities*

Interaction with the database is generally through the use of a data manipulation language such as SQL-DML [ISO(9075)87]. The minimum set of functions which must be supported are:

- store data;
- delete data;
- retrieve data;
- evaluate expressions.

As well as being able to manipulate the contents of a database directly by the use of a manipulation language, the ability to invoke DML statements from within a program is required. These may be simple commands such as those outlined above or there may be a need to apply operations to the data (e.g. addition of fields, application of simple transformations etc.). This can be provided by a macro facility, allowing procedures which perform composite operations on the data and database. An example of such a facility in an application-specific system is the Physics Analysis Workstation package (PAW) [Vandoni89].

The data may be complex, consisting of several associated data sets which are to be treated together, with the constituent parts being correctly manipulated by the operation. An important special case is that of a dataset of experimental measurements and its associated data set of the known experimental errors on each measured point. The operations on the dataset should cause the appropriate transformation to be applied to the associated error dataset. These transformations are not necessarily the same as those carried out on the dataset itself.

There is a need for the data manipulation language to be used both interactively and in batch mode. Interactive usage is exploratory, to discover the sequence of operations needed to display some particular aspect of the data; it may result in the creation of a new macro to be used in batch mode to display all subsequent datasets of that type, for example, a set of identical measurements of a parameter taken at a range of different temperatures. Alternatively, the user may want to create an animated video in batch mode, starting from an initial and final viewpoints and specifying only the endpoint and the number of steps between the two. In both cases the user merely wants to view the final screens created and stored by the batch process.

There is a well recognised need for a new level of standardization - that of database interfaces. Ideally, a "front end" application needs to be able to access data appropriate to the tasks being performed without the need to know where or how the data is physically stored. An application may require, for example, access to data which is stored in a number of physically different databases on different machines (if in a distributed environment). It is highly desirable that the interfaces to those databases should be the same, i.e. a well defined set of data definition/manipulation routines which can be utilised by the application; the mapping of those functions to actual database commands/actions being done externally. This is known as an Application Programmer's Interface (API).

4.4.2 *Archiving*

There may be a need to archive different types of data:

- original data (i.e. before any processing has been performed)
- internally derived data (i.e. having corrected for known deficiencies in the sampling instrumentation, or having performed application-specific transformations)
- pictorial or image information (officially a subset of the above)
- metadata, which is likely to be associated with the data

The formats used for each type of data may be completely different. For example, the first two may be user defined, and loaded and stored via user written routines and the third may be an ISO standard such as a CGM or a de facto one such as PostScript.

Archive technology is usually digital, but may include analogue sections, for example, a video recording of air flow associated with a jet engine in a wind tunnel may be stored together with pressure and temperature readings at specific locations within the engine.

4.5 Data Transformation

As previously stated, it is unlikely that a user will wish to analyse their original data without first having performed some form of filtering, structuring, conversion or subsetting, i.e. generating the derived data with which they wish to work.

There are many types of operation, including:

- Data Normalization
- Filtering
- Smoothing
- Grid Rezoning
- Coordinate Transformation
- Linear Transformation
- Geometric transformation
- Segmentation
- Feature detection, enhancement and extraction
- Colour table Manipulation and Feature mapping

4.5.1 *Data Normalization*

Data Normalization is applied to data (generally raw data) in order to scale it to a given range of values, for example between zero and one, thereby allowing mathematical processes or graphical presentation facilities to work with a pre-defined range of values. Normalization curves are generally linear but logarithmic or exponential curves are not infrequently used.

4.5.2 *Filtering*

Data acquired from external sources such as optical or electronic sensors, satellites etc. are inevitably affected by the sensor environment. Degradations may have been introduced as the result of sensor noise, atmospheric turbulence, camera blur (i.e. out of focus or movement) etc.. Where some known characteristic of the data (e.g. range of plausible values) is known a priori application of filtering techniques can be a useful step in "cleaning up the data". The effectiveness of a filter depends on the extent of the degradation and can often be enhanced if some information about the likely cause of the degradation is known. Filtering is widely used in the field of image restoration. [Huang79] gives a good, though dated, introduction to the use of filters in image processing.

4.5.3 *Smoothing*

When an image is recorded in a noisy visual environment (e.g. a TV image of a poor signal or a high-resolution electron micrograph), the resulting digital image may be severely corrupted by noise. A common approach to improving the visual appearance of the image is by smoothing. The most widespread technique is simply to replace each pixel by the mean of all those pixels in a region (typically 3x3) centred on it. Because the main characteristic of random noise is that it fluctuates in sign and magnitude, this has the effect of reducing the magnitude of the fluctuations.

4.5.4 *Grid Rezoning*

Grid Rezoning is the resampling of data with a structure defined by one grid to a "new" possibly overlapping, grid structure. For example, data may be mapped to a regular, rectangular grid of pixels for subsequent display. Grid rezoning is likely to be performed on derived data.

4.5.5 *Coordinate Transformation*

Sampled data may be in one coordinate system, e.g. linear but the modelling processes may require the data to be in an alternate format, perhaps because the scaling of data naturally dictates this, e.g. logarithmic or exponential, or because the physical properties associated with the data mean they are better processed or viewed in an different system, e.g. polar coordinates.

4.5.6 *Linear Transformation*

The theories of image formation are based around the ***principle of superposition***, which says that the response due to the sum of two separate signals is equal to that for a single signal of the same strength. A necessary implication of this is that isoplanatic optical imaging systems must be linear. Hence, linear transformations play an important role in the restoration and reconstruction of images. This includes topics such as deblurring (the problems of the Hubble Space Telescope have renewed interest in this area) and computerized topography, as used in body scanners.

4.5.7 *Geometric transformation*

Geometric transformations are generally used to correct data distortion (e.g. the curvature of the Earth's surface) or to generate a required perspective for viewing an object or image. There are a wide range of techniques which are encompassed by this "class". These include: projection, rescaling, rotation, polynomial warping (rubber-sheeting), translation, skewing etc.. A projection can also be considered to be a data-smoothing operation, i.e. data is never gained but often lost; projecting n dimensional data onto n-1 dimensions will generally result in the loss of structural information.

4.5.8 *Segmentation*

Segmentation has a number of different meanings, particularly when used in the context of computer graphics and visualization. However, in this context we mean the decomposition of data into a number of logical components, for example the process of partitioning a picture into a number of constituent pieces. It is generally a first stage in the syntactic analysis of an image, output being a number of primitive shapes which can be further analysed. Thresholding is a basic technique often used in the location of segments. [Mitchie85] gives a good overview of segmentation techniques.

4.5.9 *Feature Detection, Enhancement and Extraction*

Like segmentation, feature extraction (generally edges or lines) is a first stage process in image analysis. The aim of edge detection and enhancement is to locate and accentuate certain features in the data. Edge detectors are generally take the form of a mask kernel which is convolved with the image zeros or maximum values are then searched for as an indication of a "crossing" or edge. Edge enhancement does not increase the inherent information content of the data it simply emphasises specific characteristics (e.g. amplitudes in the data which represent physical phenomena).

4.5.10 *Colour Table Manipulation and Feature Mapping*

Feature mapping is used primarily for pattern recognition and scene interpretation in image processing. This is often achieved by the manipulation of colour tables (palettes). Two common techniques are:

Pseudo colouring
colours are used to enhance or assist in the location of specific features.

Histogram specification
each grey level is mapped to another by a pre-defined transformation (not necessarily linear).

[Jain89] and [Pearson91] give good reviews of techniques applicable to the field of image processing.

4.6 Data Compression

A major constraint on the exploitation of data is the volume which has to be dealt with. When large volumes of data are involved this can have severe implications for storage, memory and transmission costs. Therefore, compression techniques are very important being generally achieved either through the utilization of software methods or adoption of customised VLSI hardware.

The type of data compression algorithm employed may be dependent upon the application. Some medical or scientific applications require an exact reconstruction from the compressed file, whilst others, for example, the coding of moving video, may be able to tolerate a less exact reconstruction and generally give a greater degree of compression.

4.6.1 *Data Integrity*

There are instances where the need to retain data integrity precludes the use of compression techniques. There are two scenarios:

Lossless compression
which enables all the original image data to be recovered intact at its original precision. This is typically required for images obtained by remote sensing devices. The requirement to retain all data generally results in low compression ratios. A maxi-

mum of 50% compression is possible but less than 25% is typical for experimental data. One common technique is the Differential Pulse Code Mechanism (DPCM).

Lossy compression
which can be used when loss of precision is acceptable. This is often used when transmitting constructed images across a network and it is possible to retain the subjective quality of the image. Data storage reductions of at least 50% are common and can reach 90% or 95% with modern techniques such as fractal methods.

4.6.2 *Compression Techniques*

Although most data compression techniques are concerned with the reduction of the number of bits required to transmit or store data without an appreciable loss of information, this is not necessarily the only means by which compression may be achieved. Alternatives which generally maintain data integrity include the transmission or storage of:

- procedures by which the data may be re-generated;
- end points of vectors;
- polygonal, CSG or bi-cubic patches;
- semantic descriptions of the objects.

There are a number of different compression techniques in general use for example:

- Run length encoding (RLE)
- Entropy encoding [Pearson91]
- ADCT Adaptive Discrete Cosine Transform [Hudson87]
- PRBN Progressive Recursive Binary Nesting [Hudson87]
- IMCOMP developed by NCSA at the University of Illinois at Urbana-Champaign
- Lempel-Ziv Welch algorithm [Ziv77] and [Ziv78]
- Fractal methods [Barnsley87] and [Barnsley88].

4.6.3 *Standards*

Work to develop an ISO standard for image compression has been carried out by the Joint Photographic Experts Group, known commonly as JPEG [Wallace91]. It is not expected to become a Draft ISO Standard for at least another two years. Eventually the compression/decompression technique adopted will be implemented in both hardware and software; speculative implementations are becoming available.

Both lossless and lossy compression are described by the standard. In both cases the formats of the compressed image files will be explicitly defined and all information necessary to reconstruct the picture stored in the file, i.e they will be self-describing.

Three modes of compression are mentioned here:

Sequential mode
: images are reconstructed in one pass, from bottom left to top right. This mode is ideal for hardcopy or very fast data transmission rates.

Progressive mode
: images are reconstructed in several passes, with progressive refinement in each pass. This mode is ideal for "browsing" image databases.

Hierarchical mode
: this is still being defined and is aimed at providing a high degree of compression for very large images.

The compression factors which give equivalent subjective quality are influenced by the target display device. Hardcopy output is likely to tolerate reduced factors because, in general, the viewing distance of the paper is much less than it would be for a video display.

Recent developments in the compaction of moving pictures (MPEG) [LeGall91], fuelled by the use of related techniques in the development of High Definition Television (HDTV), are likely to provide very high compaction of computer-generated animation sequences.

4.7 Data Formats

Visualization systems need to be able to import and export data. This data may be application data (to be analysed) or pictorial information (which would require control data, graphic sequences and/or images).

4.7.1 *Generic Data Formats*

Considering the vast amounts of data being generated (or already generated), it is unrealistic to insist that all data is converted to a standard format (or formats) for visualization. It is also impractical for any specific visualization system to be expected to deal with more than a small subset of the available "standard" formats for visual data, never mind encompass the application-specific formats as well.

There have been a number of attempts to produce standard data formats, both at the low level, with relatively simple data structures, and more complex hierarchical formats. Some of the more common formats include:

Hierarchical Data Format

HDF was developed by the National Center for Supercomputer Applications, University of Illinois, at Urbana-Champaign to enable the transportation and storage of scientific data, in particular the output from supercomputer simulation results [NCSA89].

HDF is a multi-object file format for the transfer of graphical and floating-point data between different hardware platforms. FORTRAN and C calling interfaces for storing and retrieving 8- and 24-bit raster images, palettes (colour tables), scientific data and accompanying annotations have been developed. HDF allows for the self-definition of data content and purports to be extensible thereby allowing for the inclusion of future enhancements or compatibility with other standard formats.

More information on HDF is provided in "ViSC Generic Data Formats" on page 231.

Network Common Data Form

NetCDF was originally developed by Unidata for the storage and exchange of data within the space and earth science communities - [Rew90a] and [Rew90b] It is based on the earlier work on the Common Data Form (CDF).

NetCDF is a data abstraction for the storing and retrieval of scientific data, in particular multi-dimensional data. NetCDF is a distributed, machine-independent software library based upon this data abstraction which allows the creation, access and sharing of data in a form that is self-describing and network- transparent.

More information on netCDF is provided in "ViSC Generic Data Formats" on page 231.

XDR XDR was developed by Sun Microsystems. It uses the IEEE formats, and encompasses other types such as strings of characters. It is a language with a Pascal-like syntax that can describe complicated aggregates (sets) of the basic primitive element data types [Sun88].

XDR has been implemented on a variety of platforms, including SUNs, VAXs, Apple Macintoshes, IBM PCs, IBM mainframes, and CRAYs.

AVS internal format
This is an internal format for the AVS Visualization System - [Upson89a] and [vandeWettering90]. Details of AVS are given in chapter 7 which deals with Products.

apE internal format
This is an internal format for the apE Visualization System - [Anderson89] and [Dyer90]. Details of apE are given in chapter 7 which deals with Products.

4.7.2 Application-Specific Data Formats

There are literally thousands of data formats that have been adopted by different disciplines and maybe a hundred that have been extensively used on a world-wide basis. A few examples will suffice: at least fifteen more data formats were well-known to the participants at the workshop.

IEEE Formats IEEE specifies formats for the storage and correct processing of integer, fixed point and floating point numbers [IEEE85]. The formats are used in hardware by many computer manufacturers including workstations and PCs. Cray has committed itself to supporting the formats in future supercomputers. The formats are based on a precise mathematical model of computer calculation by Brown - [Coonen84] and [Brown80]. (An international standard (LCAS) is currently being defined [ISO(10967)91] for the processing of arithmetic data on computer systems - the emphasis in this case is on the arithmetic environment seen by the user, normally via a programming language interface.) Machines based on older architectures, including certain ranges from IBM and DEC, still have proprietary formats.

GRIB, GRIdded Binary form. This is a world wide meteorological format for the storage and transmission of single two-dimensional fields on a specified map projection [Stackpole89], [WMO89]. It will eventually replace existing character based codes. It can support metadata.

GF3 General Format 3 GF3 is a character based format with extensive metadata conventions developed in the UK for the transport of oceanographic and atmospheric science data [IOC84].

BUFR Binary Universal Form for Representation has been developed by the World Meteorological Organization (WMO) to represent observational type data located with respect to the earth [Stackpole89], [WMO89]. It is being used in the meteorological and oceanographical communities, and it is envisaged that it will eventually replace the currently widely used character based, telecommunication orientated formats. It can support metadata and is extensible.

FITS Flexible Image Transport System has become a standard interchange format for the astronomical community [Wells81].

Crystallographic file formats CSSR/PLUTO Crystal Structure Search and Retrieval format is used by the Cambridge Crystallographic Databank [Brown88].

4.7.3 Image/Picture Data Formats

Conceptually the simplest method of transferring pictorial information is the export and import of images. The disadvantage of using images as a means of transferring or storing data is that any information about the geometric structure of the original data is lost, (unless retrieved by difficult, relatively expensive image processing techniques). Some of the more common image formats include:

CGM The Computer Graphics Metafile has been an ISO standard (IS 8632) since 1987 [ISO(8632)87]. It has the capability to encompass both graphical and image data.

PostScript PostScript [Adobe85], or more specifically, Encapsulated PostScript Format (EPSF) is a page description language with a sophisticated text facilities. For graphics, it tends to be expensive in terms of storage compared to CGM.

GIF The Graphics Interchange Format is quite widespread, and can encode a number of separate images of differing sizes and colours [GIF87]. It also employs a powerful Lempel-Ziv Welch compression technique.

TIFF Tagged Image File Format actually encompasses a range of different formats, originally designed for interchange between electronic publishing packages. They can use several compression techniques including Lempel-Ziv Welch and FAX-like coding schemes.

PPM, PGM, PBM These formats [Poskanzer89] are distributed with the latest release of the X window system, version 11.4;

- PPM Portable Pixmap Format (24 bits per pixel);
- PGM Portable Greyscale Format (8 bits per pixel);
- PBM Portable Bitmap Format (1 bits per pixel).

XBM A one-bit image file format, standardized by the MIT X consortium.

XPM A generalization of the XBM format being developed by Groupe Bull in France.

Other formats in common usage include proprietary ones (PICT and PICT2 on Apple Macintosh; PCX on IBM PCs; VIFF within Khoros) or independent ones (FAX group 3 and 4 from CCITT, JPEG from JPEG).

At a higher level, geometric data may have been stored (generally in terms of graphical elements such as lines and polygons). Structural information is inherent and may potentially be manipulated. If this structure is purely graphical (e.g. sequences of pictures, overlays, colour maps) it is application-independent. If more sophisticated structures are transmitted, such as PHIGS structures, they are probably application-dependent, and may cause difficulties in transfer. Examples of formats which preserve information about geometry include:

- PIXAR (RenderMan) [Upstill90]
- Wavefront OBJ format
- PHIGS Archive Store [ISO/IEC(9592-1,2,3)89]
- IGES

Research into other formats suitable for describing 3D scenes is being undertaken at San Diego Supercomputer Centre (SYNU) and Pittsburgh SuperComputer Center (P3D).

There must necessarily be a trade off between image and graphical data. A typical image needs one megabyte of storage. A picture represented as a graphics sequence could contain hundreds of thousands of vectors or tens of thousands of polygons in one megabyte. Hence, except for extremely complex pictures, graphical elements are more efficient than uncompressed images for transmission.

There is a perceived requirement to transmit control information between visualization systems, but to date no attempts are known to standardize it. At present visualizations are generally exchanged via non-interactive media formats such as hard-copy or video.

4.7.4 Data Format Conversion Tools

As previously mentioned there are a myriad of different data formats available. It is probably unrealistic to expect that any one system will be able to support more than, perhaps, half a dozen of the more common ones, however with increased requirements for data (particularly images) to be exchanged, incorporated into text etc. the need to be able to transcribe or convert different formats is becoming an increasing requirement.

A number of toolkits to aid in the conversion of image formats have been developed in recent years; two of these are PBMPLUS [Poskanzer89] available on the X version 11 release 4 tape and the San Diego Supercomputer Centre image tools [Elvins90].

PBMPLUS (also known as the Extended Portable Bitmap Toolkit) is normally used in a multiplexing/demultiplexing manner, as shown in Figure 34 on page 108.

The San Diego Supercomputer Center has a suite of image tools - the Image File Format Conversion Tools - built on top of a portable library of image manipulation functions. Library functions include support for reading and writing twenty image file formats, including CGM, GIF, HDF, PIXAR, PostScript, TIFF and X11 bitmap. Once read into the image library's "Virtual Frame Buffer" (VFB) data structure, images may be resized, composited, desaturated, converted to greyscale etc.. VFBs may be recast as monochrome, colour index, or RGB images with or without colour lookup tables, alpha planes, Z-buffers, and other variants.

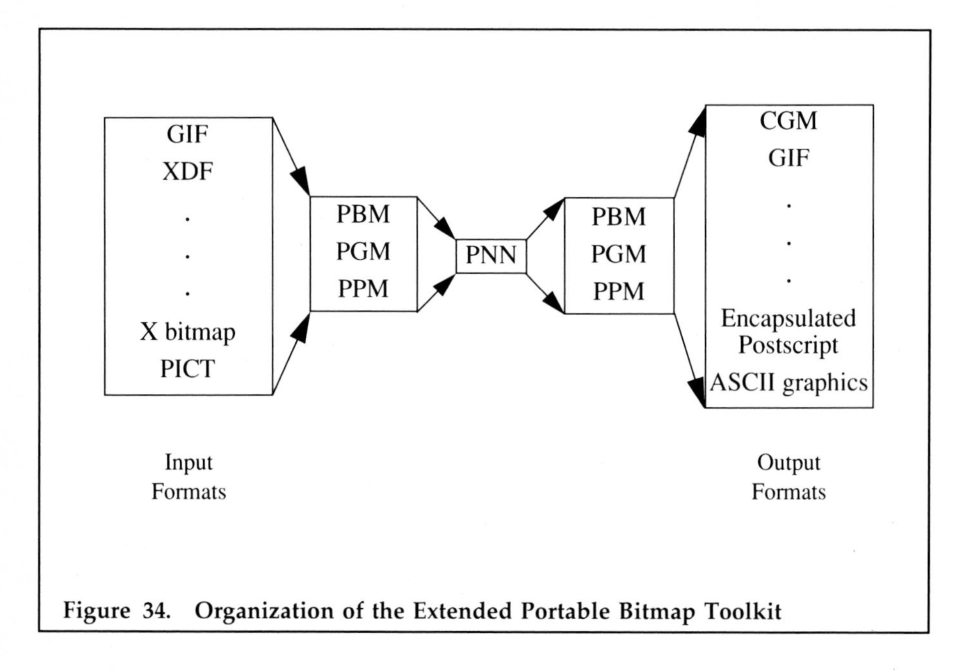

Figure 34. Organization of the Extended Portable Bitmap Toolkit

When an image is complete, it may be written out again in any of the above formats, output to a hardcopy queue via a hardcopy spooler (vpr) or displayed to a graphics screen. The tool suite is available via anonymous ftp from sdsc.edu.

4.7.5 Standards

There is a real need to develop standards for image file formats. CGM is an ISO standard but is tailored largely to the need to store vector information; at present there is no internationally recognised standard for raster data. It should be possible to define a standard for a simple image file format, which ideally would support compression. Work is underway in a number of areas, such as JPEG and MPEG. Recently IBM have submitted to ANSI for consideration a white paper defining a raster interchange format specification. This proposal does not only address raster image data but also considers multi-dimensional and non-viewable data. Encoding work is also underway for example SGML and ASN.1 are both work items in the International Standards arena. Work has begun within ISO/IEC (JTC1/SC24 Computer Graphics) to produce a standard for image processing. The current working title is "Image Processing and Interchange"[ISO91].

4.8 Recommendations

The efficient handling of data is going to be crucial for the take-up and usage of scientific visualization systems (beyond research). Ventures such as the development of HDF, netCDF and the move into scientific markets by traditional database suppliers such as INGRES, exemplify the interest in, and perceived need for, the tools and techniques for managing scientific data.

It is evident that the present plethora of data and image formats needs to be rationalised if data is to be truly portable within and between systems. It is not realistic to expect that all existing data be converted into a "chosen subset" or "standard" for input into a visualization system. What is required is that a visualization system be able to handle common data formats such as CGM, GIF, TIFF, HDF, PostScript etc., either as an integral part of the system or the application of pre-processing tools such as PBMPLUS and the SDSC tools.

Further work is needed to provide comprehensive data format conversion techniques and define standards for data transfer; ideally formats need to be self-defining. There is certainly relevant work going on in the international standards arena but in somewhat disparate groups and there is a clear need for co-ordinating and focusing of these activities.

4.9 Key References

Most references are provided in the bibliography at the end of the book. A few references which are considered key for this chapter are provided here.

[Anderson89]

Anderson H.S., Berton J.A., Carswell P.G., Dyer D.S., Faust J.T., Kempf J.L., Marshall R.E., "The animation production environment: A basis for visualization and animation of scientific data", *Technical Report, Ohio Supercomputer Graphics Project*, **(Mar 1989)**.

A technical report from the Ohio SuperComputer Graphics Project which developed apE, the animation production Environment. It provides a general overview of apE and the technical background to its design. It is necessarily outdated in respect of the latest releases.

[Bergeron89]

Bergeron R.G., Grinstein G.G., "A reference model for the visualization of multi-dimensional data", *Proc. Eurographics '89,* **(Elsevier Science Publishers B.V, 1989)**, *pp 393-399.*

Presents a reference model for the visualization of multi-dimensional data. The fundamental concepts of lattices are introduced and explored within the context of PHIGS.

[ISO(9075)87]

ISO, "Information processing systems - Database Language SQL", **(1987)**.

A formalized description of SQL, SQL-DDL and SQL-DML

[Jain89]

Jain A.K., "Fundamentals of Digital Image Processing", **(Prentice Hall International, 1989)**.

Provides a general introduction to the techniques of Image Processing

[NCSA89]

"NCSA, HDF Calling Interfaces and Utilities", *NCSA HDF Version 3.1,* **(National Center for Supercomputing Applications at the University of Illinois Urbana-Champaign, Mar 1989)**.

Gives a detailed overview of the capabilities of the NCSA Hierarchical Data Format including information on data structures, access and manipulation facilities. The manual is written from an application developers viewpoint.

[Pearson91]

"Image Processing", ed. Pearson D.E., **(McGraw-Hill, 1991)**.

Gives good reviews of techniques applicable to image processing.

[Poskanzer89]

Poskanzer J., "Extended Portable Bitmap Toolkit", **(available on X11.4 release tape from MIT, 1989)**.

Full details of the capabilities of PBMPLUS and its associated components are available in the accompanying man pages.

[Rew90b]

Rew R.K., "NetCDF User's Guide", *NCAR Technical Note NCAR/TN-334+1A,* **(Unidata Program Center, Boulder, Colorado, June 1990)**.

A User's Guide for the netCDF library of data access programs for storing and retrieving scientific data. It gives full details about the components of a NetCDF file, data types, use of the library, operations etc.. Details of C and FORTRAN bindings are supplied.

[Rosenfeld84]

Rosenfeld A., "Multiresolution Image Processing and Analysis", **(Springer-Verlag, 1984)**.

A good introduction to the field of image analysis.

[Samet88a]

Samet H., Webber R.E., "Hierarchical data structures and algorithms for computer graphics, Part i: Fundamentals", *IEEE Computer Graphics and Applications,* **(May 1988)**, *pp 48-68.*

The first part of a two part overview of the use of hierarchical data structures and algorithms in Computer Graphics. This part concentrates on the fundamentals.

[Samet88b]

Samet H., Webber R.E., "Hierarchical data structures and algorithms for computer graphics, Part ii: Advanced Applications", *IEEE Computer Graphics and Applications,* **(July 1988)**, *pp 59-75.*

The second part of a two part overview of the use of hierarchical data structures and algorithms in Computer Graphics. This part concentrates on advanced applications including octrees.

[Sun88]

"External Data Representation Standard: Protocol Specification", *Network Programming Manual,* **(Sun Microsystems Inc., 1988)**, *pp 127-142.*

Chapter 5 of this manual gives full details of the XDR protocol.

[vanderLans89]

van der Lans R.F., "The SQL standard - a complete reference", **(Prentice Hall International, 1989)**.

A brief, yet comprehensive description of the capabilities of the SQL database language. Illustrative examples are provided throughout. The book is written with the assumption that reader is familiar with the basic concepts of SQL.

Chapter 5

HUMAN-COMPUTER INTERFACE

Edited by Roger Hubbold

5.1 Introduction

This chapter deals with Human-Computer Interface (HCI) issues in relation to visualization. Following the introductory sections, the chapter is structured into two main parts: first, user issues are addressed, relating to cognition, perception, human factors and organization, then system issues are highlighted in the context of these user requirements. The chapter concludes with recommendations for future work.

A key feature of visualization is "newness", but not in concept, as the "visualization process" has been used extensively in the past and constitutes a valuable approach to scientific investigation and discovery. Increasingly, however, there is a realization that rather than just comprising a collection of novel rendering techniques, many of which are application-specific, visualization should address wider issues associated with such areas as system design, data models and HCI. This realization has arisen through both the experiences of researchers attempting to identify and tackle such topics, and via feedback from the end-users of early visualization systems. It is partly reflected in the expanded functionality available in current commercial visualization software products such as AVS and those from Wavefront Technologies.

Those with experience of existing graphics (for presentation purposes) may need convincing that there is this "newness" in visualization which can foster insight to a much greater extent than was previously possible. Visualization may modify the ways in which they appraise problems, and expand the investigative "tool box". The oft-quoted NSF report on visualization [McCormick87] defined it as a new area for research, encompassing many existing fields. User interaction is what makes visualization more than merely the union of these fields.

Visualization systems need, above all, to be flexible and extensible, so that a scientist, engineer or other user can easily construct and experiment with different ways of exploring data or hypotheses. The main components of a visualization system - hardware, software, data and user - are of equal importance and the interactions between the user and each of the remaining components should be addressed. Many existing models for visualization systems are concerned primarily with the flow of data through the system. There is a danger that any framework for visualization will reflect this bias, whereas the interaction between all parts must be seen as equally important. We need to determine whether there are different interaction characteristics associated these different parts, or whether the same rules and techniques can be applied throughout. Different aspects of the user interface which may require separate consideration are:

1. The interface for constructing the visualization "program" for a specific application. The person who constructs this interface is not necessarily the same as the end-user scientist, but will often be so. The system may be constructed via programming (using libraries) - the classical approach to computer graphics - or by means of a visual programming paradigm, or a mixture of both. In contemporary systems this is the network editor, as found in AVS and apE, described in the Products chapter.

2. Interaction with, and exploration of the model/problem/data itself, either to modify the definition of the problem, or through interactive steering of display methods and of application calculations, such as simulations.

3. Interaction with the steps in the visualization process. It may be important that the scientist can easily change the system in order to try different ways to view data. Writers of representation techniques cannot hope to predict all the different ways in which the user will want to interact either with the technique, or through it. Ideally a generic mechanism should be defined through which any external agent is able to interact with any presentation object. It is important to stress that the user must be able to modify parameters dynamically, whether of the problem definition, or some representation of it.

If, with present systems, there are different rules for these types of interaction, then we must ask: "Is there a common abstraction which would allow us to discuss interaction for visualization in a single framework, or are these inherently different paradigms which must be treated as such?"

One of the goals of visualization could be summarised as support for the graphical exploration of a user's problem. Effective exploration implies the need for a diverse range of display techniques, but the techniques must be organized sensibly to avoid overwhelming the user with options - how does the user find the closest implementation of a conceived display technique, and how is "browsing" supported - where the user is not sure how best to view a problem?

Visualization systems need to move and change in sympathy with the user's thought processes; in short, to react. We need to consider whether current interaction techniques - physical and logical - are the most efficient or natural approach for the user. Potentially the complexity of current and projected systems can result in incredibly complex user interfaces. We need to ensure that the potentially conflicting aims of power and ease of use are elegantly accommodated. Some visualization techniques become effectively unusable if the system is unable to react within a certain time. These approaches will require orders of magnitude increase in computational power before their true potential can be realized.

There is also a need to investigate the design and requirements for physical interaction devices. Forcing users to interact with their problems through non-intuitive, or even counter-intuitive media, could frustrate them and even result in the rejection of visualization systems as a problem-solving tool.

Some problems are of such complexity, or just sheer size, that the user needs interactive tools to permit even simple visualizations. Where the original data set is too large to comprehend in its original form, tools are required to facilitate the browsing and spawning of sub-problems of a manageable size. Where a problem is very complex or the user has no intuitive feel for the natural way to represent it, the system must support the ability to try different approaches almost at random, and combinations of approaches - orthogonal or lateral thinking?

5.2 User Issues

In this section we attempt to define and discuss some of the unique HCI issues in visualization, with particular emphasis on cognitive and perceptual (as opposed to system-oriented) aspects. The topics fall into the following categories:

- Cognitive Issues, concerned with mental models and problem solving;
- Perceptual Issues, such as how humans process colour, depth representations and motion;

- Human Factors, concerned with making the user feel comfortable with a system, such as representations of scale, orientation, and time, and provision of help facilities;

- Organizational Issues, related to the introduction of visualization systems into organizations.

The human mind is particularly effective at abstraction, approximation and simplification. Visualization, in the traditional sense as an internal mental process, can be characterised at this general level. The researcher may visualize the broad structure of a model, system or dataset internally, but this mental image may be simplified and incomplete in that it lacks detail and thus cannot be investigated or explained. Also, part of the rationale for visualization is that because of either the extraordinary complexity of the phenomena under study (e.g. analytically defined chaotic systems), or the sheer amount of data involved (e.g. medical imaging), or both (e.g. models of atmospheric circulation), it is simply impossible for the researcher to comprehend/manipulate more than a small portion of the model or data space internally at a given instant.

Visualization - the computing solution - can provide an alternative environment for interpretation. The visualization may be constructed as an attempt to represent the full complex model or dataset as it emerges from the (numeric or symbolic) source. Now externalised, the visualization can be:

1. graphically elaborated, i.e. rendered;

2. explored, i.e. segmented, sliced, recoloured via manipulation of a colour look-up table, rotated, or flown through.

Rendering and interaction introduce further perceptual problems relating to the complex integration of various perceptual cues associated with, for example, colour sensation, depth, motion, scale, and orientation. Some of these complexities are not unique to visualization and we may benefit from a consideration of rules and principles developed in other disciplines for the resolution of such ambiguities. Other effects, however, are new, and we have few precedents to guide us in prescribing measures to control their impact on the visualization process. Nevertheless, they must be accommodated in some systematic fashion and not regarded as insurmountable obstacles to interpretation.

Visualization systems developers and end-users stand to gain many benefits from consultation with perceptual psychologists, cognitive scientists, ergonomics experts and maybe even graphics designers and artists. As effective use of scientific visualization relies on the intellectual, perceptual and sensorial activities of the end user. In so far as current thinking in psychology offers a partial understanding of these activities, we need to be aware of how such knowledge can be usefully applied.

5.2.1 *Cognitive Issues*

For some years, cognitive scientists have been studying how users solve problems. Hayes [Hayes79] postulated that the semantics of a representation may be important in determining how difficult a problem is to solve. He showed that when subjects solve two isomorphic problems (i.e. two problems which are essentially identical in form), one after the other, there may be very little transfer between the two processes. This led to his proposal that people "fail to transfer solutions from a problem to its isomorph when the two problems differ in an important way in their semantics". He further proposed that there are (at least) six semantic categories which people use when solving problems:

- Object
- Event
- Location
- Time
- Property
- Action

Failure to solve a problem by transferring knowledge between it and an isomorph (Hayes described variations of the familiar Tower of Hanoi problem) occurs when the representation of the task is altered. In general, this changes the way that the person attempts to solve the problem - in essence, they see it as a different problem.

These kinds of issues would seem to be of central importance to the design of visualization systems, and yet little account seems to have been taken of such factors to date. For example, are the semantic categories enumerated by Hayes valuable as a way to structure concepts in a visualization system? As visualization systems can provide extensions to mental visualization processes, they should, ideally, prompt much more basic research into the methods scientists use, and into the applicability of current knowledge from cognitive science to visualization generally.

The perceptual limitations of the uninitiated user of a visualization system should not control the variety and sophistication of visualizations possible. In other words, as with any system which is expected to enable the modelling and interpretation of complex phenomena, there will be an inevitable learning curve which end users must be persuaded to travel. Potentially, the user will possess both domain knowledge and accumulated experience of a particular application, permitting the assembly of very intricate displays which, though meaningless to an outsider, may capture the relevant relationships for the user quite adequately.

5.2.2 *Perceptual Issues*

We should distinguish between the use of visualization for presenting and communicating information to others and the use of visualization in a solo context for gaining subjective insight into a large dataset or complex model. There is a huge literature on cognitive issues and a growing body of information on perceptual issues and human factors, some of which may be relevant to visualization. See, for example, [Card83], [Bruce85], [Gregory70], [Gregory77], [Shneiderman87].

5.2.2.1 *Colour*

The utilisation of colour in visualization is extremely valuable, but can be regarded as an area of particular difficulty. This reflects several factors. Although we experience the sensation of colour continuously, this experience is invariably passive and generally, we have little or no training in the active manipulation of colour.

With respect to the presentation and communication of data, many heuristics and rules of thumb exist ([Hopgood91], [Murch86], [Murch89], [Durrett87], [Shneiderman87]). However, many of these rules and warnings may become too restrictive or even irrelevant as the individual scientist starts to use the visualization tool in an investigative or analytical manner.

Empirical studies in cognitive science suggest that hue should be used for displaying nominative data (not ordinal, interval or ratio). This reflects common human experience in which we rely on a continuous stream of different colour sensations to assist us in interpreting and interacting with our environment.

Clearly, if our intention is to analyse accurately the numerical character of the data or underlying model, a scale based on hue variations alone will be inappropriate or even misleading, as hue variations do not imply magnitude. However, a scale based on brightness or saturation, both measures which have an intrinsic sense of direction, may be more appropriate. On the other hand, if our concern is to identify trends or patterns in the data, rather than explicit values, a scale involving the rapid variation that can be achieved with changing hue, may be ideal. As yet there are no widely accepted standard colour scales or range scales (such as logarithmic) for data exploration in visualization. A need is perceived for empirical studies leading to the derivation of formal guidelines and reliable standards in this area.

In general, the dimensions or axes of colour are not orthogonal. Changing one component usually causes a non-linear perceptual shift in the others. The disruption this causes to the visualization process will vary from one individual to another. Special considerations may be required for users with visual anomalies. Some research into the derivation and manipulation of perceptually linear colour spaces is already available and will be beneficial here. Perceptually linear colour spaces may also be potentially applicable to the storage, transport and reproduction of images on different platforms and hardcopy devices [Murch89].

Colour perception is clearly a subjective process and individuals differ widely in their ability to manipulate colour effectively. Potential problems relate to, for example, the incidence of colour-defective vision amongst end-users (most noticeably in males), the ease of overloading the visual system with too much colour information, and our lack of previous experience of using colour in an abstract, analytical manner.

5.2.2.2 *Illumination*

A further problem in visualization relates to the combination of lighting models and pseudo-colouring. The addition of illumination to a scene, via the numerical simulation of light sources, to produce depth-cues giving the illusion of three-dimensionality, is at odds with the need expressed above to depict numerical values and spatial relationships accurately, with pseudo-colour. This problem is particularly noticeable in still frame images of highly convoluted or non-familiar phenomena, where adding shading to imply depth alters the apparent colour of the image.

5.2.2.3 *Motion Cueing and Animation*

Fortunately, many of the visual ambiguities mentioned above are resolved when the viewpoint, light sources or objects in the scene are moved. Again, this accords closely with our everyday experience using motion to reduce visual uncertainties in perceiving our environment. Thus, we should seek, wherever possible, to construct tools for manipulating visualizations which have direct analogies in real human perceptual experience.

Visualization systems offer the potential for interpreting time-dependent systems or data, through the use of computer animation. Users should be made aware of potential problems however, such as temporal aliasing. Here, the representation of a continuous phenomenon, as a discrete sequence of still frames which run together, may introduce artefacts not present in the model. These may interfere with the perceptual process.

Also, we currently lack principles and guidelines for the manipulation of time as a variable. For example, a user may wish to accelerate time (e.g. when studying geological series data) or decelerate it (e.g. when studying subatomic processes) in the visualization, to a pace more amenable to the interpretation of the processes or model under investigation.

All the above examples highlight the necessity for configurability in the human-computer interface to visualization systems.

5.2.3 *Human Factors*

Study of the human factors of human-computer interaction is now a well-established discipline. Some, but not all, of the problems have been addressed and many guidelines have been issued [Shneiderman87]. The human factors of visualization systems, as in all systems, are essential considerations in the design. There are, however, new issues which a visualization system raises which have specific importance in its interaction with the user. Examples of considerations peculiar to visualization systems are given here, but no doubt there will be more.

5.2.3.1 *Context*

One of these considerations is the need for the user to have information available on the scale, orientation and time history of the representation being used. If the visualization enables a user to "go inside" a previously apparently solid object, all external reference points for both scale and orientation may be lost. Should this happen it is possible that, far from enhancing understanding and insight into the data, the visualization may be extremely misleading. Similarly, in a display using animation of images, there could be a need for information on elapsed time or rate of change of time in the display. Without a time reference the value of the display may be diminished.

The scale, orientation and time data can be classified as metadata on the visualization. Metadata are necessary to set the display in context and should be required output from all systems. The use of multiple active simultaneous views is also provided by many visualization systems and context labelling of the views is particularly necessary for the user of this facility.

5.2.3.2 *Help and Documentation*

Many visualization systems are very large and, as in any large system, context-sensitive on-line help and good hard-copy documentation are essential. The sensitivity to context of a help facility must specifically include the context of visualization. For example, help provided by text alone may not be as appropriate for problems in image manipulation as help provided by a visual representation or a pictorial example such as a reference image. The visualization facilities provided by the system should, where possible, be used by its own help facility. In addition, the help should ideally be user configurable. The need for documentation to be on paper alone may impose severe restrictions on its effectiveness. Video and other more appropriate representations will need to be employed, and should be as carefully planned as documentation text.

5.2.4 *Organizational Issues*

The introduction of large software systems into organizations raises a host of issues which cannot be ignored - [Eason88] presents these issues for IT in general, using the author's experience of ergonomics. Visualization systems are just such types of systems and these organizational issues should be addressed early in the planning for their introduction. They include such things as re-allocation of resources from other areas to support the system; support and training for users; effects on inter-departmental relationships of introducing a visualization system in just one department; and ownership of data between departments.

One of these issues, which may be peculiar to a visualization system, is its effect on interactions with colleagues. The user of a visualization system probably works alone at the screen-face, but is not alone when working as a member of a project team or part of an organization. The introduction and presence of a visualization system may well have an effect not only on the organization, the project teams and the users' colleagues, but also on the users' ways of thinking. Visualization enables better understanding of, and insight into, data by the user. This understanding will need to be communicated to colleagues. Just as it is likely that visualization will change ways of scientific thought because of the availability of more advanced tools, so it may also change ways of communicating by making new channels available. For example, an instrument designer, who needs to explain requirements to an engineer, can use a 3D interactive, rotatable image to display an idea rather than a set of 2D cross-sectional plans. Thus, the availability of a system may alter the communication of scientific understanding between colleagues.

The potential value of a visualization system to an organization could be immense, because of the enhancements it provides to users' understanding and insight. The potential advantages are, unfortunately, balanced by potential problems. Many of these problems can be avoided by careful management and planning. It is possible that in scientific research establishments, which are likely candidate organizations for visualization installations, the opportunity for education in understanding the system and in management of the problems, may be constrained. This would be to the disadvantage of individual users, of organizations and of the scientific community as a whole.

5.3 System Issues

In this section we examine hardware and software issues relating to support of the human-computer interface. Many current systems can be seen to be "technology-driven" - the way that a user interacts with a system is determined, even constrained, by the underlying workstation hardware and software. In the longer term it is important that the needs of users of visualization systems lead to improvements. We live in hope!

5.3.1 *Flexibility*

The need for flexibility in visualization systems stems from the requirement to be able to explore alternatives. Sometimes it will be possible to achieve this by putting together existing software components, whilst at other times new code will need writing. Thus, the user interface part of a visualization system will require:

1. Toolkits which support common forms of interaction, such as menus, scrollbars, and other typical window manager tools. These tools already exist, but are not necessarily implemented in a way which matches all requirements of the user interface. For example, the choice of an underlying graphics or window system may severely constrain the incorporation of new types of physical device.

2. Tried and tested tools for displaying data and exploring it. Simple examples include isosurface tilers, data slicers and colour map editors. Note, however, that these current examples can be seen as parameterised filters. They do not deal with techniques for direct manipulation of the user's model.

3. Methods for adapting existing tools - in effect, programming by example. The flexibility which this implies is seen as crucial to the central role of visualization as a means to facilitate experimentation.

The use of a visual programming interface which allows standard tools to be plugged together, exemplified by AVS and apE, is one current approach to this. Some kind of "folding editor" which permits the internal details of modules to be viewed and modified would allow the creation of new modules from existing ones. In many respects, this is akin to the inheritance of objects and methods in object-oriented systems like Smalltalk, but we specifically do not intend to imply that an object-oriented system should necessarily form the basis for a visualization framework.

5.3.2 *Separation of Logical and Physical Models*

It is important that the physical and logical models of a user interface are separated. This implies a sound logical framework which can be mapped to real devices, for both output and input. The major motivation for this is portability - we need systems which run on a variety of platforms, but which can also be tailored to exploit local features, such as novel input devices, where these are available.

5.3.3 *Virtual and Physical Interaction Devices*

We propose that devices should be divided into two classes: physical and virtual.

5.3.3.1 *Physical Devices*

These are real hardware devices, such as mice, keyboards and dials, together with more recent innovations such as data gloves and N-space cubes. It may prove to be important that a visualization system supports other kinds of input as well, such as speech, TV cameras or other kinds of sensor, or kinaesthetic devices (e.g. force-feedback devices). This will require a more general framework for input handling than is offered by current graphics or window systems.

5.3.3.2 *Virtual Devices*

At one time this name was used for what are now called logical devices in systems such as GKS and PHIGS. We mean something different. We have purloined the term to describe higher-level tools, constructed from one or more physical devices, and (possibly) other virtual devices. A fundamental attribute of these virtual devices is that they should be **user-programmable**. This permits application-specific feedback to be attached to them. Three examples are:

- A "sound probe", which can be moved around in a model space, and which emits sounds according to the data encountered at that location. Bergeron and Grinstein have described such a tool [Bergeron89].

- A constrained cursor which moves on some surface within the model space.

- A "weather ruler" which could be used to measure parameters, such as pressures or wind velocities, within a numerical weather model.

This kind of semantic feedback is important in visualization - it embodies the notion that the user is probing the application space, rather than only its visual representation (although the latter may well be regarded as a valid part of the model). The idea of a data probe, which provides application-specific feedback, is quite important: the scientist will usually need to see detailed quantitative results as well as overall relationships.

This style of interaction is termed direct manipulation. A major issue which this raises is whether the user interface can be separated from the application. Virtual devices which have application-specific feedback cannot be divorced from the application. They entail invoking application code as part of the feedback loop. Systems which divorce the user interface from the application may prove inappropriate for this kind of interaction.

5.3.4 *Dialogue Management*

A visualization system should be able to support multiple control threads. This implies that it should be capable of responding to external (or internal) stimuli asynchronously.

Typically, event-driven systems provide this capability, but the precise method of implementation has a fundamental effect on the granularity of the asynchronism. This is a thorny issue because it is dependent on the operating system and the graphics/window system, but it does affect the ease of use of the system.

To give a very simple example: it should be possible for a user to interrupt (in a clean way) some process which has been activated, such as a request to render a large volume, which the user decides to abandon. With some systems this may be difficult to achieve. In particular, the X Window System does not provide a convenient mechanism for supporting multiple threads within a dialogue, unless they correspond to different (UNIX) processes. It is possible that reliance on the current X Window System is undesirable in the longer term.

There are several approaches to dialogue management which are widely used - [TOG86] and [Foley90a]. These have different characteristics and some may be significantly better than others at providing the flexibility needed by a visualization system. This is an area which requires further study. Examples are:

State transition networks - in this approach, the dialogue is represented by a directed graph in which the nodes represent states and the arcs represent transitions between states. Arcs are followed in response to events caused by user input or other system actions. In some respects, state transition networks are similar to data flow networks used in some visualization systems. They tend to be unwieldy for representing multiple control threads.

Grammar-based techniques - a set of rules determines the structure of the dialogue and the actions which process an input stream. Examples include BNF grammars, multi-party grammars, and production rules. By representing actions with tokens it is possible to mix different kinds of input, such as speech and graphics.

Event-driven systems - in these systems, a set of functions must be provided to handle input events from different devices. Synthetic events can be constructed to handle composite devices, or higher-level devices.

Examples include NeWS [Gosling89], which employs PostScript procedures to handle events distributed by a central distributor, and X11, which places events in an event queue and distributes them on demand. In X11 toolkits, a call-back mechanism is employed whereby an application-specific function can be invoked when an event from a particular tool is removed from the event queue.

These approaches have different strengths and weaknesses and cannot be easily mixed together. Further studies are required to determine the styles of interaction which would be useful for visualization users.

Such studies would also be useful for assessing existing products. For example, AVS handles input by attaching devices to specific modules. It is unclear how this fits with the concept of virtual devices which are user-programmable, or how conveniently a single event from an input device can be distributed within an AVS-like system to multiple modules.

5.3.5 *Context Sensitive Help*

Visualization must be able to support context-sensitive help. This may take many forms, including textual information and visual clues. The system must maintain enough state information to give **intelligent** assistance. Additionally, systems should be adaptable to different users' native languages.

5.4 Conclusions and Recommendations

As visualization can provide extensions to mental processes, it should prompt more basic research into the methods scientists use to solve problems. Specifically, work is required to see how knowledge from cognitive science can be applied to the design of visualization systems.

A large body of knowledge about user interfaces exists. It is important to include users' requirements and to build tools which use guidelines wherever possible. For example, some current systems break known guidelines by having very complex menus, which are difficult to navigate unless one is quite familiar with them. These will be off-putting to the occasional (but perhaps creative!) scientist. Similarly, some current systems have tiny menus and other tools which are difficult to control because they do not take account of Fitt's Law [Card83].

Colour is another example where many people using visualization systems make mistakes, such as using inappropriate colour scales, or mixing colour and shading cues. In addition, the ability to specify colour information in an unambiguous, device-independent manner is vital to the transfer of visualization images between display media and also assists the reverse mapping from picture to data in some cases. Empirical studies are needed in order to formulate guidelines and reliable standards in this area.

Visualization tools should build on users' everyday experience of things such as interpretation of motion, thereby capitalizing on real human perceptual experience.

Visualization systems can be very large and complex. It is important that they provide visual clues to a user regarding parameters such as scale, orientation and time. Equally important is the provision of context-sensitive help to guide the user through the system.

The potential benefits of visualization within organizations could be immense, both as an aid to problem solving, and for communicating knowledge to others. But to obtain these benefits requires careful planning for the changes which such systems can bring about.

New styles of interaction are important and attention should be paid to these. For example, multi-threaded dialogues may be needed, so that the user can interact simultaneously with different parts of a model. Studies are required to find what kinds of interaction are most useful for visualization, and to devise software systems which can support these.

The integration of new devices will be important. In order to maintain portability of systems, there should be a separation between physical and virtual devices. The incorporation of new devices into current window and graphics systems may present problems.

The use of highly interactive computer graphics coupled with novel input/output technologies, our lack of a complete understanding of human perception and our unfamiliarity with many of the kinds of systems and data studied within visualization bring major new HCI problems for which there are few precedents. Extensive consultation with cognitive scientists and the close involvement of end users in the system design process are required to tackle these issues.

5.5 Key References

Most references are provided in the bibliography at the end of the book. A few references which are considered key for this chapter are provided here.

[Bergeron89]

Bergeron R.G., Grinstein G.G., "A reference model for the visualization of multi-dimensional data", *Proc. Eurographics '89*, **(Elsevier Science Publishers B.V, 1989)**, *pp 393-399.*

In addition to providing a reference framework for the visualization of multidimensional data, the paper describes the use of icons for representing multiple dimensions, seven in the example given. The proposed framework supports sound as well as graphics.

[Bruce85]

Bruce V., Green P., "Visual Perception: Physiology, Psychology and Ecology", **(Lawrence Erlbaum Associates, 1985)**.

Contains a large amount of detail on perceptual issues, including many interesting illustrative examples.

[Card83]

Card S.K., Moran T.P., Newell A., "The Psychology of Human-Computer Interaction", **(Lawrence Erlbaum Associates, 1983)**.

A standard reference work covering various aspects of modelling users, including perceptual, cognitive and motor processing, and the role of short and long term memory. Also contains practical information about modelling users' performance, such as the keystroke model for predicting efficiency of dialogues involving typing.

[Durrett87]

"Colour and the Computer", ed. Durrett J.H., **(Academic Press, 1987)**.

A useful book covering most of the important topics in colour, including perception, colour models, hardcopy devices, and descriptions of various applications which capitalize on colour.

[Foley90a]

Foley J.D., van Dam A., Feiner S.K., Hughes J.F., "Computer Graphics, Principles and Practice", **(Addison-Wesley, 1990)**.

Updated issue of the earlier standard reference book on computer graphics by the two principal authors. This new edition is usefully expanded and has good material dealing with user interfaces at an introductory level. It also shows how software for interaction can be integrated with graphics software.

[Gregory70]

Gregory R.L., "The Intelligent Eye", **(Weidenfeld and Nicholson, 1970)**.

Useful as a general interest book on how we see the world, and how we learn to see it and interpret it.

[Gregory77]

Gregory R.L., "Eye and Brain: The Psychology of Seeing", **(Weidenfeld and Nicholson, 1977)**.

Another general interest book, well worth reading for background information. Has interesting material on use of stereoscopic images, and drawing in three dimensions.

[Hayes79]

Hayes J., "Cognitive Psychology and Interaction", *Methodology of Interaction*, ed. Guedj R.A., ten Hagen P.J.W., Hopgood F.R.A., Tucker H.A., Duce D.A, ISBN 0-444-85479-7, **(Elsevier North-Holland, 1979)**.

This is an interesting paper which deals with how users formulate solutions to problems. In particular it postulates that users tend to describe and think about problems in terms of specific kinds of objects and actions. These are described using the familiar Towers of Hanoi problem, and the paper shows how changing the problem description may make it hard, or even impossible for people to solve mentally.

[MacDonald90]

MacDonald L.W., "Using Colour Effectively in Displays for Computer-Human Interface", *Displays*, **(July 1990)**, *pp 129-141*.

Provides a useful introduction to using colour on displays for the human-computer interface, including the use of perceptually uniform colour spaces. It provides a set of guidelines for the use of colour, but warns against applying them too rigidly. A large number of references on the subject are provided.

[Monk84]

"Fundamentals of Human-Computer Interaction", ed. Monk A., **(Academic Press, 1984)**.

A collection of papers dealing with visual perception, human memory, thinking and reasoning, collection and evaluation of behavioural data, workstation design, user interface design, speech systems, and human factors.

[Murch86]

Murch G.M., "Human Factors of Colour Displays", *Advances in Computer Graphics*, ed. Hopgood F.R.A, Hubbold R.J., Duce D.A., **(Springer-Verlag, 1986)**.

A tutorial covering perception of colour and showing the derivation of rules for practical use of colour.

[Murch89]

Murch G.M., "Colour in Computer Graphics: Manipulating and Matching Colour", *Advances in Computer Graphics*, ed. Purgathofer W., Schoenhut J., **(Springer-Verlag, 1989)**.

An update on [Murch86] dealing with the manipulation of colour using various colour models and perceptually linear colour spaces. Also highlights the problems of matching colours on different devices.

[Robertson86]

Robertson P.K., O'Callaghan J.F., "The Generation of Colour Sequences for Univariate and Bivariate Mapping", *IEEE Computer Graphics and Applications* vol 6 (1), **(Feb 1986)**, *pp 24-32.*

This paper advocates some principles to be followed when using colour to represent the value of one or two system variables in an ordered manner.

[Robertson88]

Robertson P., "Visualizing Colour Spaces : A User Interface for the Effective Use of Perceptual Colour in Data Displays", *IEEE Computer Graphics and Applications* vol 8 (5), **(September 1988)**, *pp 50-64.*

Deals with the use of perceptually uniform colour spaces for displaying data. The paper describes the application of the CIELUV and CIELAB models and includes useful practical examples.

[Shneiderman87]

Shneiderman B., "Designing the User Interface", **(Addison-Wesley, 1987)**.

Contains a large number of guidelines for designing user interfaces, including those dealing with cognitive issues, through to use of graphic design principles and use of colour for effective presentation.

[TOG86]

"Special Issue on User Interface Design", *ACM Transactions on Graphics* vol 5(2), 5(3), 5(4), **(1986)**.

These three special issues cover software for user interfaces, including user interface management systems, styles of interaction, object-oriented systems, direct manipulation and window systems.

Chapter 6

APPLICATIONS

Edited by Peter Quarendon

6.1 Introduction

This chapter gives some examples which illustrate practical applications of visualization methods and experiences with them in a variety of fields. It shows which methods are being applied to the solution of current problems and the motivation for their use. It also points out some of the difficulties which have been experienced. Thus it shows which areas are being well catered for by visualization methods and the areas where it is felt that gaps and shortcomings exist and where more development is still needed.

In some areas, such as molecular graphics, visualization is well established. From these, useful parallels can be drawn to suggest the direction which the newer areas may evolve. Where new techniques have recently been developed or where data visualization is being newly applied, readers may not be aware of developments and may be stimulated themselves to try graphical methods.

6.2 Chapter Structure

Visualization is used to some degree in almost every area of science and engineering. The choice of examples here strongly reflects the interests and expertise of the workshop attendees, but an attempt has been made to select applications with as wide a spread of subject area and to cover as many different types of visualization as possible. In particular, there are examples from areas where the data is gathered from the real world, and examples where the data is computer generated.

Examples of both two-dimensional, three-dimensional and four-dimensional data are included. In an attempt to draw together those which have similar visualization requirements, the applications have been categorised according to the type of their principal data sets, using the classification scheme explained in chapter 3 (Visualization Techniques).

The classes considered are:

E_2^{nS}

A number of data values are associated with each position in a two-dimensional space.

- Cartography
- Study of Statistical Indicators

E_3^S

A single scalar value is associated with each position in a three-dimensional space or, equivalently, a continuum of data values is associated with each position in a two-dimensional space.

- Remote Sensing

E_3^{nS}

Data values are associated with positions in a three-dimensional space.

- Analysis of Archaeological Data
- Physical Chemistry and Drug Design
- Biochemistry
- Materials Research
- Medical Science
- Archaeological Reconstruction

$E_{2;t}^{nS}$

Data values are associated with positions principally in a two-dimensional space together with a series of time steps.

- Meteorology
- Ice Stream Visualization

Although both these examples do have three spatial dimensions, one, the height, is less extensive than the first two and many visualizations are two-dimensional.

$E_{3;t}^{nS}$

Data values are associated with positions in a three-dimensional space and a series of time steps.

- Oceanography
- Oil Reservoir Engineering
- Computational Fluid Dynamics

E_m^{nS}

Data values are associated with positions in a higher-dimensional space

- Dynamics of Systems
- Program Visualization

This does not fully reflect the range of application areas covered by the workshop participants, but does cover most of the types of dataset.

Each section covers a single subject area and is structured as follows:

1. An introductory description, giving the background and motivation for the use of data visualization in the area.

2. A description of the input data, that is the main data sets forming input to the data visualization process. These are typically the output from a simulation or from measuring devices after any standard processing, such as Fourier inversion, has been carried out. The dimensionality of the data, values recorded and typical sizes are noted.

3. A summary of the visualization processes typically applied to the data:

   a. Derived data

      This is data which is derived from the primary input data to assist in its understanding. Examples are cross-sections, equivalued surfaces or the positions of maxima and minima. These serve the purpose of showing interesting features of the original or to reduce its dimensionality.

   b. Visualization techniques

      The methods of display which have been, or are expected to be, found useful in understanding the data.

4. As in the available space it is not possible to give more than a brief summary of the application, key references for further information are also noted.

6.3 Cartography

Although cartography is recognised to have existed for hundreds of years (millions of maps providing the evidence) maps are commonly regarded as being presentational rather than analytic research tools. This assumption is erroneous. While no-one can deny the value of maps in presenting characterised abstractions of spatial (normally earth related) features, the analytical scope of a map is powerful and has been used by scientists (e.g. climatologists, physicists and epidemiologists as well as geographers) for hundreds of years. The key is that maps offer direct visual analysis and comparisons of the spatial relationships between wide ranges of separate distributions. Despite their remarkable successes in the past, paper maps with their dated content and restricted scope for data storage, are limited tools for the study of massive datasets and complex models. The development of the computer map with related attribute datasets in Geometric Information Systems (GIS) however has changed this radically. Although such systems are currently being used extensively for land and utilities management, they already have proven power in analysis. As such they could be regarded as scientific visualization systems primarily for "geographical" datasets/models. This example is based on a GIS application in which the bringing-together of various spatially related data can reveal spatial patterns which can provide scientific insight.

A project which illustrates this integration of remotely sensed data into a geographical information system is the Islay Case Study. It aims to show the potential of GIS for conservation and resource management, through the ability of GIS to integrate disparate data sources and produce new data layers through that integration. The derivation of new data layers, and the techniques which enable this, show GIS to be more than a tool of presentation graphics. Techniques such as polygon overlap, line and point buffering allow data of varying numbers of dimensions to be integrated. The ARC/INFO Geographic Information System product (from ESRI) was the main software used for the project.

Input Data

A number of different data layers were integrated in the project. The base layers came from three separate sources, namely:

1. Conventional Cartography - data on soil, forestry, rivers, roads and tracks were prepared in digital form by manual digitisation from existing paper-based maps. The data form a number of separate layers, which are then used for a variety of different analyses. These data sets tend to be large, for example the data coverage for the topography of Islay contains spot heights and contours, numbering around 17,000 data points.

2. Field Survey - very detailed information on items such as the location of nesting sites of rare bird species are not available from sources other than field survey. The location and attributes of such information was then prepared in digital form by either digitisation of field maps, or through manual entry of attribute information.

3. Remote Sensing - multi-spectral satellite imagery from LANDSAT Thematic Mapper was classified using the GEMstone image processing system and then transferred to the GIS for further analysis. Aerial photographs were a further source of data and digital data was obtained from these by the manual digitising of interpreted photographs.

Visualization

A number of different data layers were derived from the initial input data. The simplest sets of derived data come from selections of particular spatial features depending on their attributes in the non-spatial database, or on topological relationships with other spatial features. For example selecting all roads which were "A" class, and which passed through areas of forestry. The topological relations can be those of containment, e.g. select all points with given attributes which are within polygons of another given attribute. Alternatively the relations could be of proximity, e.g. select all areas of forestry which are next to peat bogs.

The simplest visualization method involved shading the spatial data according to particular attributes in the non-spatial database, for example, according to land use classification. More advanced visualization involves the construction of digital elevation models, where a triangulated irregular network of the topography of the area is constructed and then viewed in perspective. Rendering of the surface is carried out using a simple "fish-net" grid. In addition, other categories of data can be draped over the basic terrain diagram, enabling previously unseen relationships to be appreciated. For example, it is possible to drape the land use map over the topography, to begin to examine the relationship between the two.

A demonstration version of the system was constructed using the ARC MACRO LANGUAGE (AML) running under DECwindows. This demonstrator allows the user to interrogate the data in a variety of ways, without knowing the nature of the database query language. Inter-relationships between the various data elements can be visually investigated in this way.

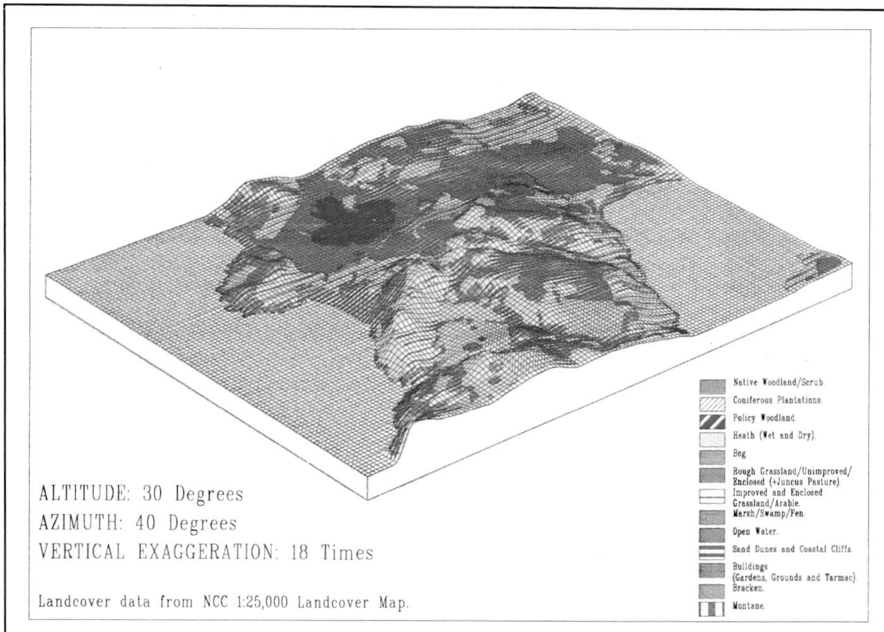

Figure 35. Perspective view showing Landcover.: Thirteen different types of landcover are superimposed on a vertically exaggerated map. Both colour and hatch style are used to distinguish the areas. By this means the geographer seeks to summarise masses of observations and to explore relationships discernible from the data. *Terrain model digitised by Diane Brown from Ordnance Survey 1:25000 map sheets; picture compiled by Tim Rideout, Project Computing Officer, using Arc/Info; supplied by David Mitchell, Department of Geography, University of Edinburgh; Crown Copyright reserved.*

6.4 Study of Statistical Indicators

This example is concerned with the problem of the definition of the statistical indicators used in thematic mapping and in social area analysis. These indicators attempt to quantify abstract concepts, such as poverty or market potential, which are difficult to define precisely let alone measure. These diagnostic indicators have to be derived from an array of descriptive data on related measurable phenomena. For example, data on male unemployment is often used as a primitive indicator of poverty or low market potential when targeting areas for positive discrimination or product promotion.

Visual explorations of alternative representations of the same data revealed that conventional ratio measures are highly misleading indicators of the distribution of many phenomena. Other metrics portrayed the distributions more reliably and were more attuned to policy requirements. This lead to the use of the signed chi-squared measure instead of the traditional ratio measure for mapping of high resolution census data.

Input Data

For illustrative purposes, the example is limited to a set of two variables, namely the number of males in the economically active age groups for inhabited one-kilometre grid square areas.

Visualization

In this research, a set of displays are used with data in tabular form. These alternative views of the same data are produced as and when necessary to pursue lines of enquiry. Examples are:

- distribution maps against a base map of the number of males unemployed and the percentage of males unemployed;
- scatterplots of percentage of unemployed against the number of unemployed males;
- geographic maps of the distribution of male unemployment as indicated by various metrics, including the signed chi-square measure;
- Lorenz curves for active and unemployed males respectively for each of the measures being evaluated (see Figure 36 on page 140).

This is a type of visual exploration not facilitated by current systems. It emphasises the need for the interactive use of a set of maps. All the thematic displays, including tables, provide alternative visualizations of the underlying phenomena. Thus manipulating elements on one display involves the manipulation of the underlying data model, which should produce corresponding effects on other displays. The set of displays and the nature of the interaction can be variable and unpredictable and need to be configured at run-time to mirror the trend of thought and mental visualizations of the researcher seeking an insight.

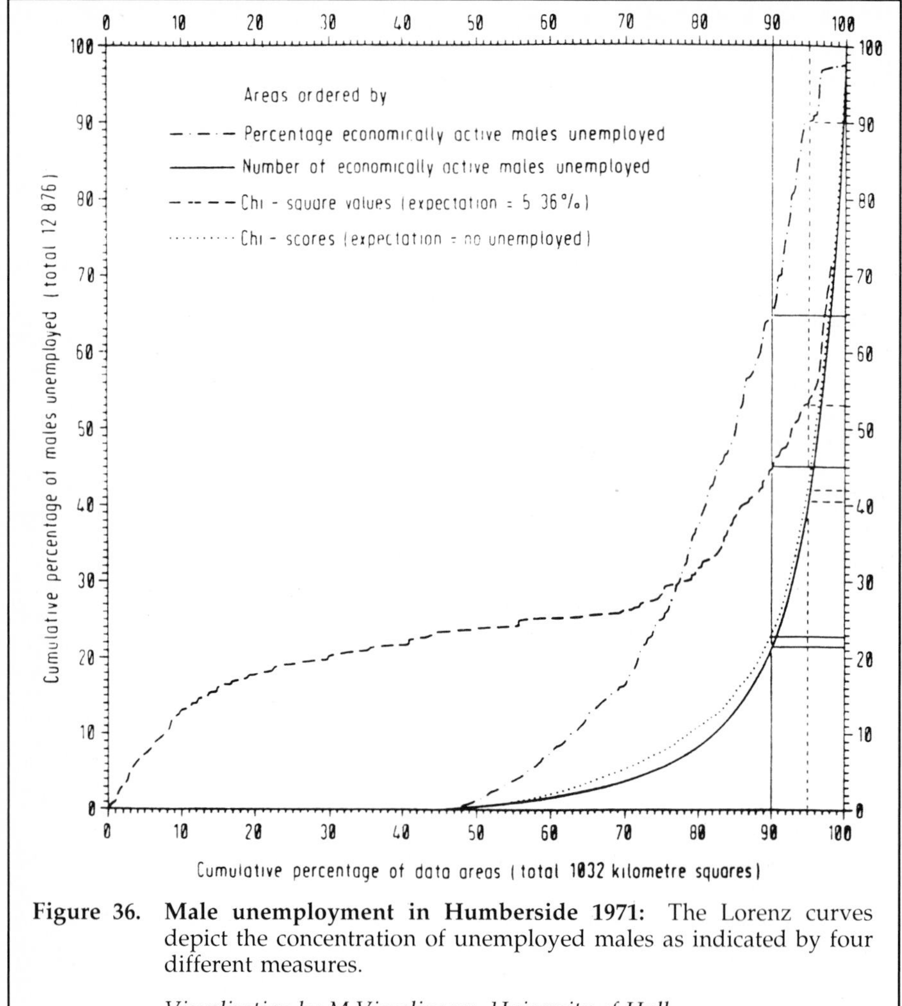

Figure 36. Male unemployment in Humberside 1971: The Lorenz curves depict the concentration of unemployed males as indicated by four different measures.

Visualization by M.Visvalingam, University of Hull.

References

[HMSO80]

HMSO, "People in Britain - a Census Atlas", **(Census Research Unit/ Office of Population Censuses and Surveys/ General Registrars Office (Scotland), 1980)**.

Contains the distribution maps for census data.

[Visvalingam81]

Visvalingam M, "The signed chi-score measure for the classification and mapping of polychotomous data", *Cartographic Journal* vol 18 (1), **(1981)**, *pp 32-43*.

Describes the development of the measure.

6.5 Remote Sensing

Satellite and airborne imaging devices are producing increasing quantities of data, particularly from imaging spectrometers. One motivation is to be able to identify and estimate ground features, such as specific types of vegetation, more easily.

Input Data

A typical modern imaging method (e.g. from AVIRIS) produces data in 200 bands in the visible and near infra-red frequency range. Long image strips are generated, up to 16,000 pixels wide. This can be considered a field with two spatial (but not quite Cartesian) dimensions and one of frequency.

Data from radar altimeters produces height data at a similar resolution.

Visualization

The analysis of remote imaging has a long history and a wealth of two-dimensional image processing methods have been developed. However as the number of frequency bands has increased, new methods are being adopted to allow for the increased dimensionality of the data. Reported visualization methods display horizontal (two dimensions of space at a particular frequency) and vertical (one dimension of space and one of frequency) sections through the data. The emphasis is on interactive examination of the very large data sets. Typically, two such sections are displayed simultaneously and in each a line displays the intersection with the other. The section being displayed can be changed simply by moving the line with a mouse. There have also been visualizations of three-dimensional terrains constructed from aerial and satellite data. One such, from LANDSAT data, is shown in Figure 37 on page 142.

Figure 37. **Three-dimensional visualization from satellite data:** The height information is obtained from digital terrain maps; the colour is derived from the thematic mapper. *Visualization by the Jet Propulsion Laboratory, California*

References

[Hibbard89]

Hibbard W., Santek D., "Interactivity is the Key", *Proc. Chapel Hill Workshop on Volume Visualisation*, **(May 1989)**, *pp 39-43.*

Describes a system for analysing multi-spectral satellite data.

[Kenada89]

Kenada K., Kato F., Nakamae E., Nishita T., "Three dimensional Terrain Modelling and Display for Environmental assessment", *Computer Graphics* vol 23 (3), **(1989)**, *pp 207-214.*

Shows the type of results which can be obtained from mapping aerial photographs onto a terrain model.

6.6 Analysis of Archaeological Data

Modern day archaeology is not just confined to time consuming and expensive excavations but also other forms of field works such as aerial reconnaissance, geophysical prospecting, collecting artefacts from plough-soil, and topographic surveys. This type of field work produce vast amount of two- and three-dimensional data which can only be analysed using computers. Advanced data visualization techniques have offered a great opportunity to the archaeological community to obtain a clear understanding of their collected data.

Input Data

Input data are usually three-dimensional positional data and other types of two-dimensional survey data, such as resistivity data, magnetometer data and ground radar. Devices such as resistivity probes, electronic distance measures (EDMs) and digitisers are used to collect those data. Typical resolution for a resistivity or magnetometer survey might be on a 100 by 100 grid over an archaeological site.

Large quantities of sample data are also collected, in some cases tens of thousands of individually positioned data samples have been recorded in three dimensions.

Visualization

Input data from sensing devices is usually processed by image processing methods to provide a smoother surface or to remove noise.

Survey data is, by its nature, scattered and often needs to be interpolated so that it can be combined with other data for display.

The data is then typically displayed as wire-frames and scatter plots in two and three dimensions. Wire frame pictures, showing possible effects of ploughing on the distribution of pottery fragments are shown in Figure 38 on page 144.

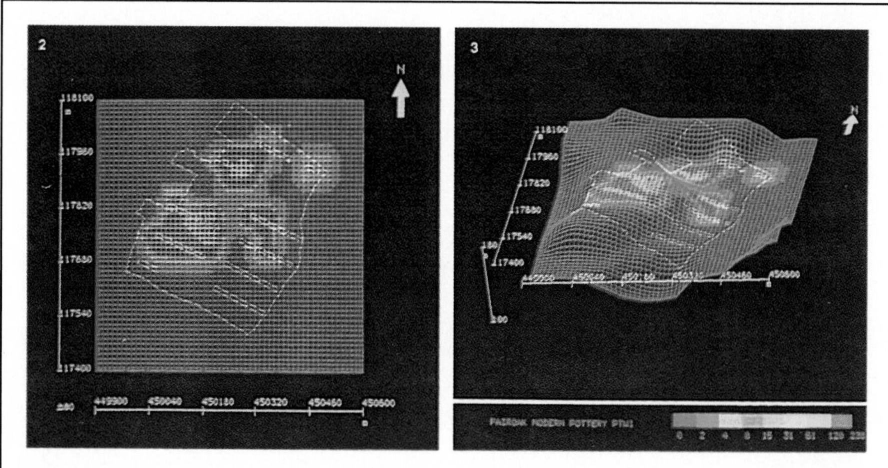

Figure 38. **Pottery distribution at Fairoak:** The pictures suggest how pottery fragments may move through the action of ploughing. *Data courtesy of Hampshire County Council. Visualization by the IBM UK Scientific Centre.*

Reference

[Reilly89]

Reilly P., "Data Visualization in Archaeology", *IBM System Journal* vol 28 (4), **(1989)**.

Gives a spectrum of examples of visualization used to analyse archaeological data.

6.7 Physical Chemistry and Drug Design

An aim of physical chemistry is to predict accurately the properties of small molecules - a necessary precursor to the computer design of new drugs, and so predicting the effect of new compounds without having to synthesise and try them. Some simulation codes are empirical, while others start from the basic wave equations. Ab-initio calculations are accurate but very time-consuming - being measured in days - and can only be attempted on simple molecules.

Input Data

The principal output data from a simulation are the exact positions of the atoms in, and the electronic structure of, the molecule. The format of the latter is somewhat peculiar to the type of simulation but often consists of partial charges at the atomic positions. This can be reduced by sampling to a series of scalar values at regular grid positions in the space around the molecule. Typically, a single scalar value such as the electrostatic potential is computed on a 100 by 100 by 100 grid.

Visualization

Figure 39. Properties of small molecules (a): The hydrophobicity of a small molecule (analapril). *Data courtesy of Dr W.G.Richards, Oxford University. Visualization by the IBM UK Scientific Centre.*

From the basic atomic positions, the atomic surface is derived. The most common is the ***van-der-Waals surface***, formed simply by representing each atom type by a sphere of standard radius centred at the atomic position. The spheres of bonded atoms intersect and the resulting model gives an approximation to the volume and shape of the molecule. Other surfaces such as those accessible to a solvent molecule are also used. Although the atomic properties are field values defined everywhere, they are usually only of interest at or near the molecular surface. Thus the space-filling model is often colour coded according to a property such as its electrostatic potential or affinity with water (hydrophobicity is the inverse of this property). Figure 39 on page 145 shows the hydrophobicity colour-coded on a small segment of an analapril molecule.

Figure 40. Properties of small molecules (b): The electrostatic field round the same molecule (analapril) is shown by a series of equipotential surfaces. *Data courtesy of Dr W.G.Richards, Oxford University. Visualization by the IBM UK Scientific Centre.*

Alternatively, chosen equipotential surfaces are displayed, as illustrated in Figure 40.

Often, a surface model is augmented by a stylised model of the molecular structure, as coloured balls at atomic positions and lines (or sticks) showing the bond positions. This shows the position and orientation of the molecule in a clear way, for example when investigating the interaction of two molecules.

Many other visualizations have been developed for special purposes, including the following.

Interaction between molecules

The interface between two interacting molecules can be defined by taking the surface which is equidistant from the closest point in each molecule, the "mid-interface" surface. This surface can be coloured by properties relating to the interaction - for instance the difference in electrostatic potential.

Comparison of molecules

A representation which is useful in this context is the "gnomonic projection". This reduces the dimensionality of the data by projecting the field value at the molecular surface onto a surrounding sphere. Comparisons between molecules can then be made by comparing the fields on these two-dimensional spherical surfaces.

References

Any issue of the *Journal of Molecular Graphics* will provide numerous examples of the use of visualization in molecular chemistry.

6.8 Biochemistry

Biochemistry has a long history of use of data visualization methods, particularly in the study of large molecules such as proteins. Proteins have complex three-dimensional structures which are the key to their activity. Physical models of such molecules, which usually consist of many thousands of atoms, are fragile and very difficult to work with and chemists have long been obliged to work with computer models to achieve their results.

Molecular graphics in biochemistry appears relatively mature. The Journal of Molecular Graphics, whose principal subject matter is visualization methods in molecular chemistry, is now in its eighth year of publication and activity in new representations and displays appears to have slowed in recent years. Further, in contrast with most other subject areas, three-dimensional (and four-dimensional) graphics are in regular and routine use.

Input Data

A typical task is the determination of the structure of a protein, now carried out on a routine basis. For such large molecules, computational methods cannot yet be used to determine the three-dimensional shape. Instead, crystallographers use X-ray diffraction to construct a map giving the density of electrons at each point in the three-dimensional space of the molecule when in its crystalline state. Typically, such a map is computed on a 50 by 50 by 50 grid, each grid interval corresponding to about 1 Angstrom.

Also, by chemical methods, it is possible to obtain chemical formula of the protein, in the form of the sequence of (perhaps 200) amino acids in the molecular chain.

Visualization

Electron density maps are usually converted to equipotential surfaces. The contour levels are chosen to correspond with relevant features (for example six membered rings) in the data. Ridge-line plots can also be used. The atomic structure is represented by a ***bond plot***, which represents each inter-atomic bond by a line. Often, the line is coloured in two halves to show the type of atom at each end.

The aim of the process is to arrange the atomic positions of the known protein structure so that it best fits the electron density. A portion of a molecule after this fitting is illustrated in Figure 41 on page 149. This is achieved by adjusting the atomic structure interactively so that the bond lines follow the appropriate electron density levels. Automatic energy minimization methods are used to refine the hand fitted data.

The complexity of many displays makes transparent surfaces a necessity. Most interactive work is done using surfaces composed of regularly spaced dots or grids rather than continuous shading. Strong depth-cueing, the use of controllable clipping planes and stereo viewing aids are considered necessities to reduce visual clutter.

Reference

[Morffew85]

Morffew A.J., "Protein Modelling using Computer Graphics", *Advances in Biotechnological Processes 5,* **(1985)**, *pp 31-58.*

This gives a good, if now somewhat old, general description of molecular representations.

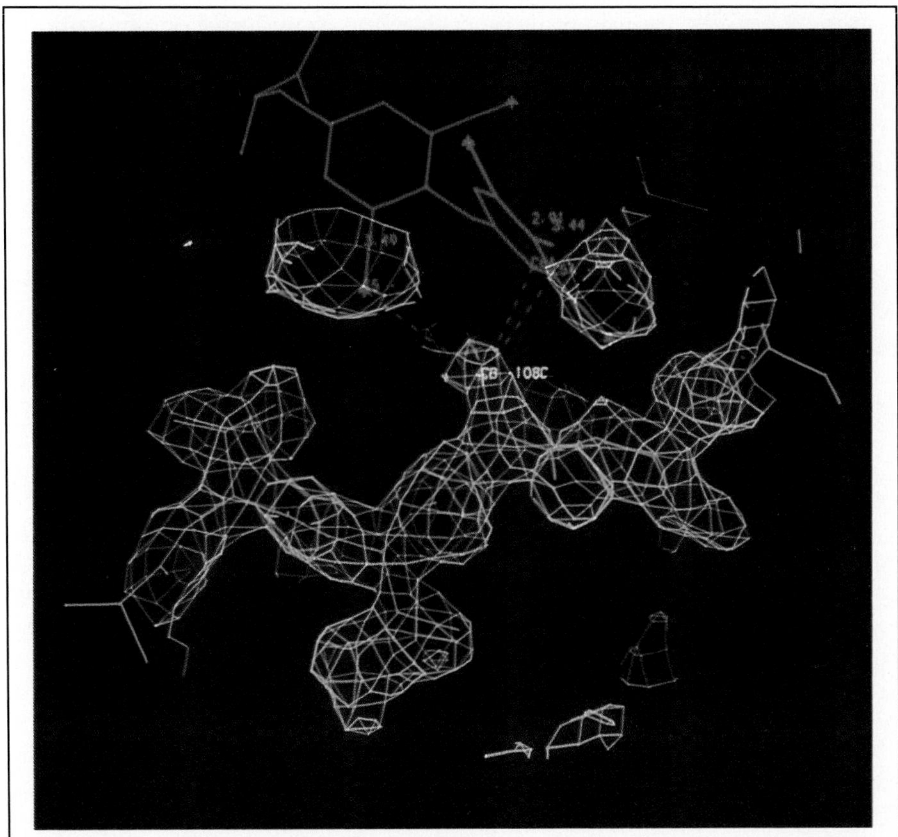

Figure 41. Region of a protein molecule: A small portion of the protein dihydrofolate reductase is shown as a green bond plot. A relevant equipotential surface is shown in blue. *Visualization by Dr. J M Burridge.*

6.9 Materials Research

Materials research, or condensed matter research, is concerned with the determination of the structure of materials. The structures are found by subjecting the material to a particle beam and analysing the patterns in the scattered particles. The particles may be photons, electrons or neutrons. The experiments are done over a range of temperatures, pressures and magnetisations. The materials may be solids or liquids. Solids may be crystalline, powdered or amorphous. There is increasing interest in the properties of thin films and layers of differing materials. For instance, the electronics industry needs to measure semi-conducting materials; chemists want to understand catalysts such as zeolites, or properties of surfactants. Also ceramics and high temperature superconductors are becoming increasingly important industrially.

Computer simulations are also performed on large numbers of atoms. Many unit cells are replicated and the scattering properties of the ensemble calculated for comparison with experiment.

Input Data

These typically consist of counts from particle detectors for photons or neutrons. There may be thousands of detectors, they may be pulsed at repetition rates of 50Hz or higher and the time dependence of the data can be important. Newer instruments have area detectors which collect much larger data sets. Electrons usually form 2D images, but recently many 2D slices have been taken to form 3D datasets.

Visualization

During data collection there is a need for fast multi-dimensional viewing of 3D spatial views, overlaid with other measured parameters, shown as a function of time.

At later stages, it is necessary to display refined atomic and molecular positions. In solids these are relative to the unit cell, in liquids they are compared with simulations. Molecular graphics are needed for individual molecules, but more often for arrays of molecules, showing how individual ones pack to form complete structures. Figure 42 on page 151 shows the unit cell of a zeolite: Sodium zeolite A.

The greatest inhibitor to the successful use of graphics is that suppliers seem to be unaware of the "Standard Crystallographic File

Structure" for the interchange of data between molecular graphics and analysis packages.

Reference

[JMG85]
"Standard Crystallographic File Structures-84", *Journal Molecular Graphics* vol 3 (2), **(June 85)**, *pp 40.*

Defines the proposed Standard Crystallographic File Structures.

Figure 42. Structure of sodium zeolite A: The structure is illustrated for a single unit cell, the different structural elements being shown by different geometric shapes. *Visualization by Dr. B H Collins, IBM UK Scientific Centre.*

6.10 Medical Science

Many diagnostic procedures in medicine now produce three-dimensional data sets.

Present practice is for the radiologist to look at a series of two-dimensional cross-sections and to build up a mental picture of the three-dimensional structure. They become highly skilled at this but it is thought that benefits would be derived if three-dimensional structures could be displayed and manipulated directly. Two applications of interest are:

- Planning surgery, for instance in orthodontic and cosmetic procedures. Here the three-dimensional situation is often very complex and it can be far from clear what the results of a proposed modification might be. The hope is that this could be accurately modelled.
- Planning radiation treatment. The problem here is to find the optimum orientation and placement of radiation beams to maximise the effect on a tumour while minimizing the effect on surrounding organs.

At present, little clinical use is evident in this country, although widespread in the USA.

Input Data

Computer Tomography (CT) scans use X-ray sources and a series of detectors to produce a map of the opacity to X-rays across a body slice. Taking a series of these slices gives full three-dimensional data. Bone structures show up well in CT images but only a limited number of slices are normally taken because of the dangers of large X-ray doses. For diagnostic purposes, a CT scan might consist of 25 slices, each perhaps 256 one byte values (pixels) square. (From cadavers, much longer scans are taken.) The distance between slices is usually much greater than the distance between pixels in a slice.

Magnetic Resonance Imaging (MRI) uses a strong magnetic field to excite hydrogen atoms and, by noting the decay from various directions, is used to produce a map of their density across body slices. Many soft tissue structures show up well with this method. Because no adverse effects of the high magnetic fields have been detected, larger numbers of slices can be taken with MRI, up to 100 by 512 by 512.

Positron Emission Tomography (PET) gives good data on areas of physiological activity, but has poor localisation. Ultrasound also produces three-dimensional data-sets.

Multi-modal medical visualization [Hu89] seeks to merge functional information from different sources. For example PET data on cerebral metabolic activity may be merged with anatomical landmarks - for example, brain surface morphology - from CT or MRI.

Visualization

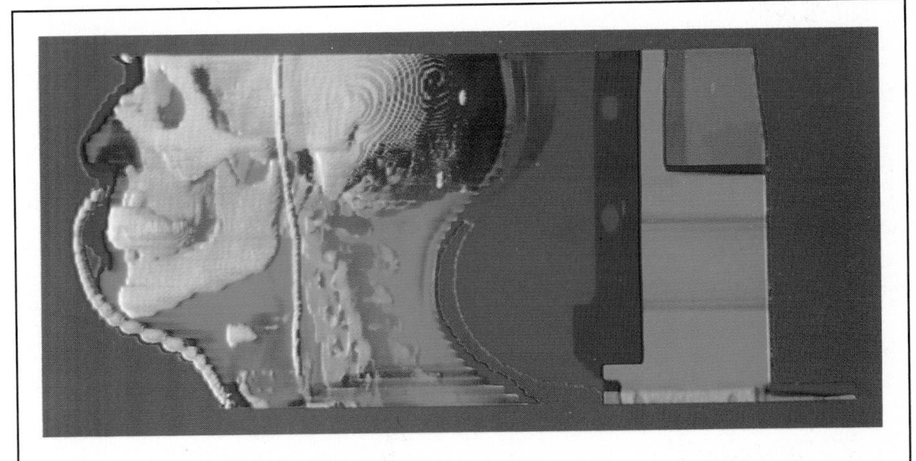

Figure 43. Three-dimensional reconstruction from a CT scan: The image depicts a reconstruction of the skull, facial bone structure and upper spine of a male subject. The colour and opacity of the various materials displayed are: air/skin/fat - red, semi-transparent; soft tissue - green, also semi-transparent; bone - grey, opaque. It is believed that the Moiré artefacts on the skull are related to the steepness of the ramps used in classifying the various materials distinctly from each other. The blurring around the mouth occurs partly because of scattering caused by fillings in the teeth and partly because of a restraint which the patient is asked to bite on during the scan, in order to prevent excessive head movement. Part of the scanning bed is visible on the right.

In spite of the skill of the radiologists one very useful facility is to be able to display a section through the region of interest in an arbitrary direction. This allows a radiologist to verify, for example, that an intended direction along which to reach a particular site during an operation is in fact reasonable. Ideally, this would be interactive. Sectioning can be done without interpreting the data, an important consideration ethically.

The principal process which is applied before three-dimensional display is the labelling of regions according to the type of material present.

Identifying the tissue and bone on a CT scan can be done by straightforward thresholding, as shown in Figure 43 on page 153. MRI data is less easy to label mechanically. At present, for example, brain regions in a skull have to be identified in each slice by a human operator. The direction of research seems to be to provide aids to this process, for instance with automatic boundary tracking, rather than to completely automate it.

Having labelled regions, full three-dimensional renderings can be made of the structures, using cut-away techniques or transparency to reveal underlying shapes and their relationships. Because of the differing resolutions within and between slices, some interpolation is necessary if resulting three-dimensional reconstructions are to look smooth.

Reference

[Fuchs89a]

Fuchs H., Levoy M., Pizer S.M., "Interactive Visualization of 3D Medical Data", *IEEE Computer*, **(Aug 1989)**, *pp 46-51*.

This paper gives typical examples from one of the most active groups in the field.

6.11 Archaeological Reconstruction

The solid modelling of archaeological reconstructions, using systems designed for engineering have become a topic of increasing interest in recent years. Archaeologists recognise the value of the three-dimensional model as, perhaps, the most rigorous test of any theories concerning long-vanished sets of buildings. The models have, in turn, become one of the most compelling ways of presenting the interpretation to the public.

The optimum use of this type of reconstruction is where the computer model is defined from the beginning of the interpretative process - where it can be used as a tool in its own right. The model can then be elaborated and refined as the evidence itself is built up and, in a final stage, may be prepared in a suitable rendered form for interpretation.

Input Data

Excavated data reveals the shape of the building. Therefore input data is geometrical for this type of applications. These geometrical shapes are specified either as Boundary-representation data or as Set-theoretic data, depending on the type of modeller being used. In set theoretic terms, a typical data volume might be 20,000 planar half-spaces for a single building. In many instances, the desire is to model a complete site such as Pompeii. In such cases, models can be much larger. Associated data gives the colour, texture and other attributes of the parts.

Visualization

Figure 44. Archaeological visualization: A model of the Chapter House, Kirkstall Abbey near Leeds. *Visualization by the Department of Computer Science, Leeds.*

If the input data is in Boundary-representation form, polygonal data is derived for the purpose of visualization. However, if the input data is set theoretic, there are renderers which operate directly from the set-theoretic model. In either case, the data structures are rendered by scan-line methods, by radiosity or by ray tracing to produce realistic images. Figure 44 on page 155 shows a view of the inside of an early abbey.

To expose the structure of complex models, it is often necessary to add sectioning planes and other cut-aways.

As the purpose is to give an impression of the environment as it may have been, interactive (even "virtual reality") walks-through are sought. The data-sets are very large, however, and sufficient performance cannot be obtained from present technology. Animations have therefore to be prepared by assembling many still pictures onto video-tape.

Reference

[Reilly90]

"Communication in Archaeology : a global view of the impact of information technology Volume One : Data Visualization", *Proc. Second World Archaeological Conference*, ed. Reilly P., Rahtz S., **(July 1990)**.

Shows the current interest in and gives recent examples of data visualization in archaeology.

6.12 Meteorology

Everybody is affected by the weather and climate and scientists have long tried to understand and hence predict changes in both weather and climate. Numerical weather prediction was first attempted before 1922, but only became practical with the widespread availability of computers in the 1950s. Climate prediction has only been seriously attempted in the last few years. Previously efforts concentrated on simulating the observed features of the existing climate, but now current computing power is allowing attempts to predict the climate assuming various boundary conditions. Models of smaller features of the atmosphere, such as hurricanes and individual thunderstorms, (mesoscale, as opposed to synoptic scale) or even smaller, at the micrometeorological scale, such as airflow within an agricultural crop are also used.

Data Visualization is essential to allow meteorologists to assimilate large quantities of complex multivariate data, often in a very short time.

Conceptual models of the dynamic behaviour of the atmosphere are reasonably well formed, though there is still active mathematical research into the existence of analytical solutions of various differential equations. The interaction of the chemical constituents of the atmosphere are understood at a qualitative level. Research now concentrates on refining existing models, both conceptual and numerical, quantitatively improving basic data sources and solving problems in reasonable time. To paraphrase the oft quoted Richard Hamming: ***"The purpose of computing is insight, THEN numbers."***

Generally, the numerical models are large and complicated, and are the products of teams of people working for many years. Climate experiments may take many months of supercomputer time to complete, but weather forecasting models are run repeatedly every day to very tight schedules (a rough rule of thumb is that one can afford to spend n hours on producing and disseminating a 10n hour forecast).

Input Data

The primary source of data is a vast network of instruments across the globe. 7000 measuring temperature, humidity, pressure, wind speed and direction at the surface, 700 measuring these throughout the depth of the atmosphere, and numerous instruments on ships, aircraft and satellites. Other variables are also measured. Frequency of measurement is from 12 hourly through to every few minutes, with most at least hourly. Most of the data is circulated globally as rapidly as possible. Much of the data is processed at the instruments to produce more significant values (e.g. wind speed is averaged over 10 minutes).

Derived data consists of a previous numerical forecast, which is used to interpolate the observational data, either statistically ("optimum interpolation") or using the numerical model to achieve an interpolation that is dynamically consistent spatially and temporally ("data assimilation"). The observational data are interpolated onto a regular grid (or analysed into spherical harmonics) and transformed into other variables that are not easy to observe directly. These form the initial conditions for numerical prediction. At present satellite imagery is only used qualitatively as current computer power is insufficient to incorporate it into models correctly.

Visualization

The complexity of meteorological data and the long history of manual methods has produced a unique display method.

The raw data are plotted, using an internationally agreed symbolic notation, onto maps for inspection, usually many variables at once. A single variable may then be "analysed" to produce a contour map. Quality control is performed on the data during the analysis.

These maps are produced for different levels in the atmosphere, and are often superimposed. Differential air flow at different levels is a strong predictor of subsequent atmospheric evolution and is a more accurate indicator of vertical air motion.

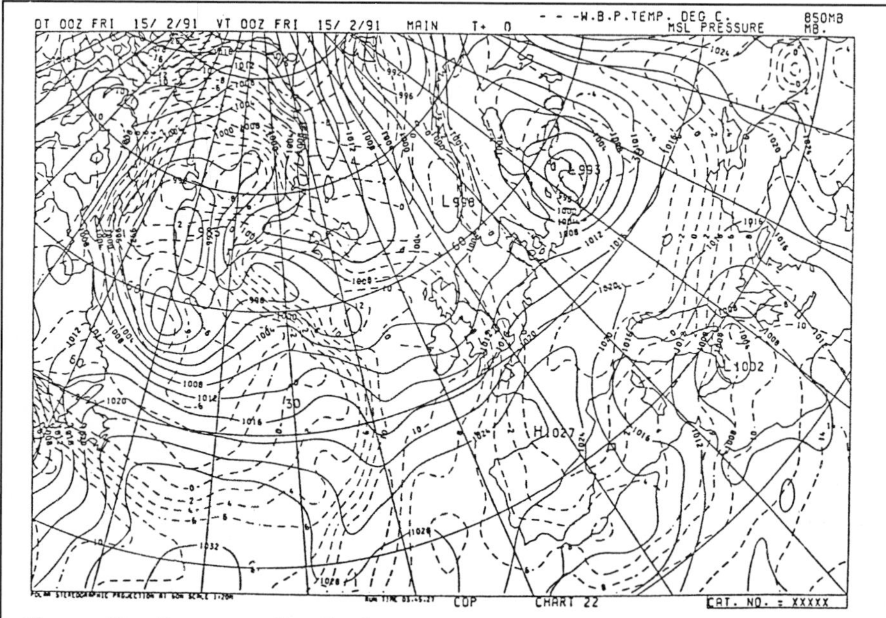

Figure 45. Pressure distributions: A meteorological computer-generated chart of analysed surface data ('analysed' - original observational data, quality controlled and interpolated to a regular grid). Mean sea level isobars are overlaid onto a form of isotherms (actually 'wet bulb potential temperature' - a temperature taking into account moisture content and varying pressure).

The maps are always superimposed on topographical outlines to allow local orographic effects, that cannot be simulated adequately, to be estimated. Fixed scale charts are used to allow overlaying and easier estimation of quantitative values.

The maps at the surface level often have discontinuities (fronts) incorporated. Computer techniques are still not adequate to draw these satisfactorily.

In mid latitudes, mass (pressure) and wind velocity are intimately related, so that a pressure map can act as a wind map and vice versa. In the tropics, stream-line charts must be used. Trajectory plots are used for pollution studies.

Specialized vertical cross-sections, to highlight the thermodynamic structure of the atmosphere are also used.

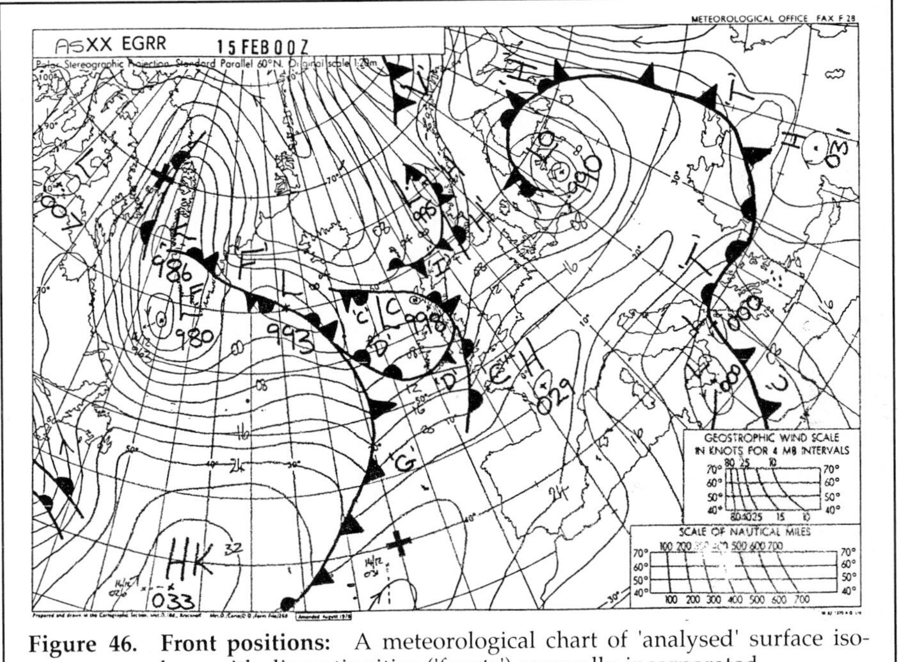

Figure 46. Front positions: A meteorological chart of 'analysed' surface isobars with discontinuities ('fronts') manually incorporated.

Originally the use of colour was restricted by the technology. Now only limited use of colour is made, ensuring that monochrome charts, which are much cheaper to produce and reproduce, are still useful. Contour lines at standard values are used in preference to continuous shading for ready extraction of quantitative information. Shading is usually reserved for more qualitative features such as rain and fog.

Interaction is limited, usually to select the next picture or combination of overlays. Observations are often selected ("picked") during quality control for modification or deletion after comparison to other data.

References

[Little87]

Little C.T., "Graphics at the UK Met Office", *Proc. BCS Conference on The future of graphics software*, ed. Earnshaw R.A., **(1987)**.

Examples of meteorological graphics, for computer scientists.

[MO87]

Meteorological Office, "The Storm of 15/16 Oct 1987", *Met Office Report*, ISBN 0861-80-2322.

Many examples of meteorological graphics, for meteorologists.

[Seum89]

Seum C.S., Wilcox R.W., "Aspects of AMIGAS II Design and Implementation", *Proc. Second ECMWF Workshop on Meteorological Operational Systems*, **(Dec 1989)**, *pp 121-125.*

Describes an example of a meteorologically orientated visualization system.

6.13 Ice Stream Visualization

Ice streams are areas of fast moving ice within ice sheets. A stream has the potential to remove a large amount of ice cap material over a short period. The British Antarctic Survey (BAS) have a finite element model of the Rutford Ice Stream which is being used to study the behaviour of the stream under conditions of global climatic change. Visualization provides a useful tool in this area to animate the changes in the ice stream modelled over a run representing many thousands of years.

Input Data

The finite element mesh used by BAS contains nearly 700 nodes covering the complete ice stream. Each node is a vertical triangular prism which contains data describing ice thickness, temperature values in the vertical section and state information showing whether the ice is floating or grounded. Geometric information is also present describing the position of each node. During a model run the state of the ice stream is sampled at regular intervals yielding, in the example used, 200 time steps.

Visualization

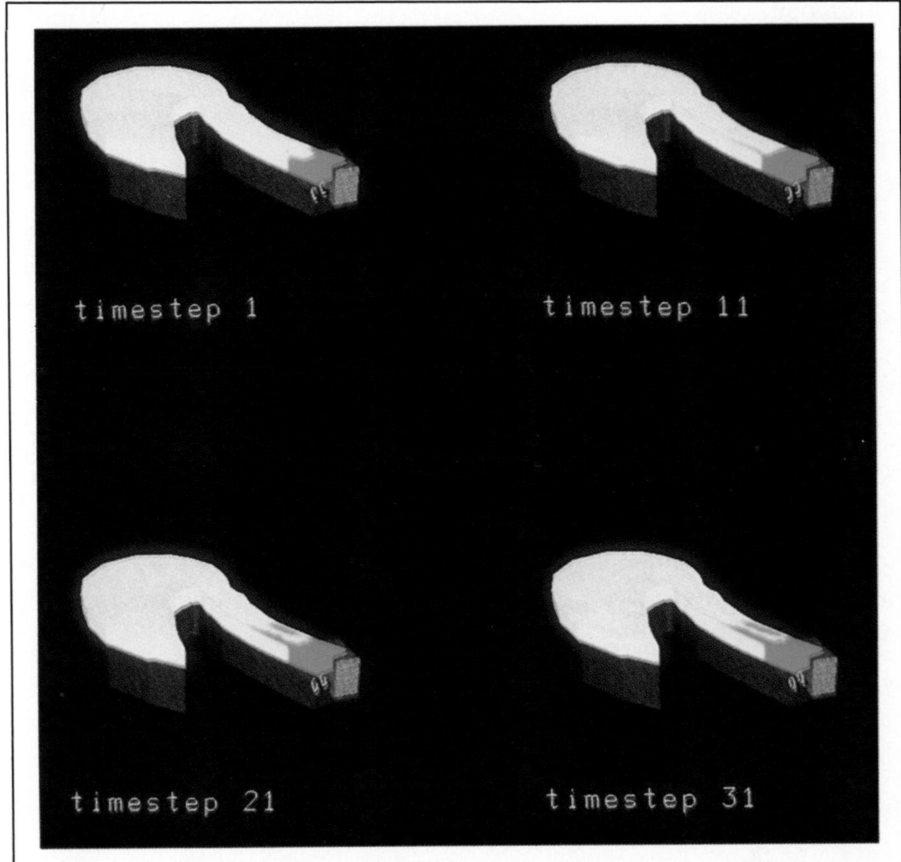

Figure 47. The Rutford Ice Stream: Grounded ice is shown in white and floating ice in green. *Data courtesy of the British Antarctic Survey; visualization by the IBM UK Scientific Centre.*

The input data geometrical information is used to construct a prism for each element. The prism is a primitive solid object in WINSOM, the set-theoretic modeller which is used. Each prism is colour coded to show whether the ice is floating or grounded. A high quality 24 bit shaded image is generated for each of the 200 time steps produced by the model. These images are then copied to video tape and the resulting animation shows the progress of floating ice upstream in the ice stream in response to rising sea level. The use of simulated time has greatly increased the understanding of the nature of this migration of floating ice.

References

[Frolich89]

Frolich R., "The Shelf Life of Antarctic Ice", *New Scientist*, **(Nov 4, 1989)**, *pp 62-65.*

Describes the behaviour of the ice flows being studied.

[Quarendon84]

Quarendon P., "WINSOM User's Guide", *IBM UKSC report 124*, **(1984)**.

Documents the WINSOM solid modelling system.

6.14 Oceanography

An example of current oceanographic research is the Fine Resolution Antarctic model being used to model the Southern Ocean. One of its aims is to verify or disprove the global warming hypothesis.

The simulation model computes many variables including temperature, salinity, vorticity and is capable of resolving energy transport eddies of approximately 100Km.

Visualization of the temperature and vorticity fields (one scalar and one vector) is vital for understanding the behaviour of the resulting flows.

Input Data

Data values are stored as horizontal sections through the ocean with a vertical resolution of 100 metres. The horizontal resolution is about 10 kilometres. The total size of one data set is greater than 1 gigabyte and new datasets for further time-steps are created several times each week. The model has now reached 16 years into its projected 20 year lifetime.

Visualization

Temperature data (covering the range -0.5 to +29 degrees C) is compressed into two byte signed integers or one byte unsigned characters depending on the application tool used. Stream function data may be derived from velocity, salinity and density data.

One characteristic of the temperature data which is noteworthy is that relatively small temperature excursions are superimposed on a large latitudinal temperature gradient. Because of the limited dynamic range of the visualization techniques, this latter may have to be removed before sufficient detail can be observed.

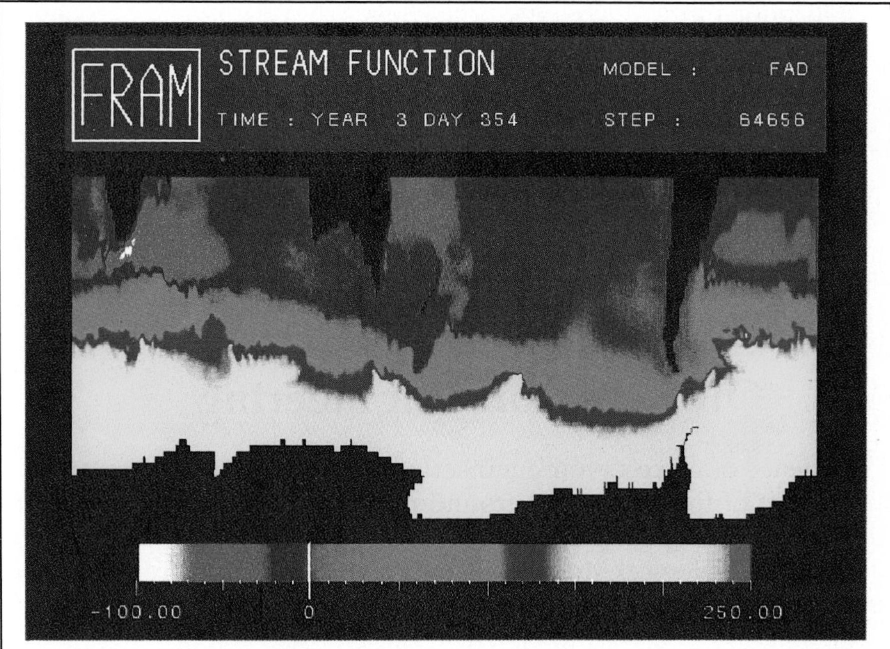

Figure 48. Stream Function Display of Circum-Polar Current: The intensity of the Circum-Polar current is mapped in false colour in a pseudo-Mercator projection.

Graphical data analysis is only just beginning for this model. Various methods are being tried, including isosurfaces and direct volume rendering by ray casting to show the internal structures. Colour-mapped sections are also expected to be used and methods of data overlay are to be investigated.

No simple method has yet been found to give ideal results, but the work is continuing.

It is anticipated that interactive movement of cutting planes and the ability to interactively change the colour mapping will be vital to bring out the fine detail in the variation of properties such as temperature. Control of transparency is also needed in the volume rendering methods.

The overall data-set is four-dimensional, showing the complex movement of the ocean, the formation and evolution of eddies and so on. These can only be appreciated by animating sequences of similar visualizations from successive time steps.

Reference

[Cox84]

Cox M.D., "A Primitive Equation: 3-Dimensional Model of the Ocean", *Geophysical Fluid Dynamics Laboratory Ocean Group Technical Report 1*, **(Princeton, 1984)**.

This describes the FRAM model.

6.15 Oil Reservoir Engineering

The science of oil reservoir engineering deals with the occurrence and movement of fluids in underground reservoirs and their recovery. In recent years, complex mathematical analysis of the behaviour of fluids within the geological structures and the increased use of computers has lead to greater recovery. Even so, it is not uncommon for 60% of the total oil in a field to be left in the ground.

Oil and gas are becoming more costly to find. Furthermore, some of the new finds are in areas of very complex geology. The oil companies are therefore having to use more sophisticated techniques to analyse the behaviour of reservoirs.

Input Data

The reservoir grid is an irregular 3D grid and may contain voids (inactive cells) and faults (adjacent cells not able to have cross-flow). Also, areas of large activity such as that around a well bore may use locally refined grids and areas of little activity may be amalgamated into larger grid blocks.

Many values may be associated with each grid block, including rock properties, the amount and type of fluid in place, its pressure etc.. Connectivity between blocks via the wells may also need to be shown.

Visualization

It is important to be able to generate new data from existing data in a general way and also to extract data.

The main displays used are:

1. Line graphs and histograms to show, for example, the variation of pressure with time or the proportions of various hydrocarbons in particular regions.

2. Contour displays to display, for example, regions with high oil content at a particular time.

3. Bounded regions to show the value of one or more variables, perhaps the amount of oil, gas and water, in each grid block.

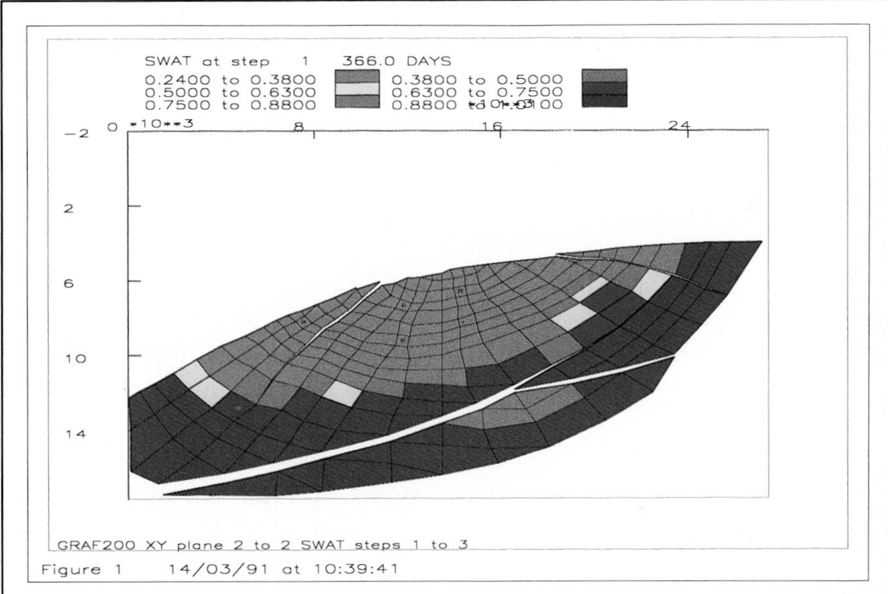

Figure 49. Oil reservoir simulation: The picture shows a horizontal section through the reservoir data from one time step in a simulation. *Data and visualization by INTERA ECL Petroleum Technologies.*

Figure 49 shows the estimated amount of water on a horizonatal section through a reservoir at a single time step in the simulation.

Reference

[Watkins87]

Watkins H.K., "Graphics in Reservoir Simulation", *Computer Graphics Forum* vol 6, **(1987)**, *pp 111-118.*

This gives further information on the use of visualization in oil reservoir exploration.

6.16 Computational Fluid Dynamics

Visualization has long been applied to experimental fluid dynamics but is relatively new in computational fluid dynamics (CFD).

Usually, CFD produces large data sets at the end of long batch jobs, often undertaken by supercomputers. The role of visualization in CFD is, as Gary Belie [Belie86] put it:

> *Too much output data from a supercomputer could swamp the analyst. But workstations and custom software let engineers look into - and through - complex flow systems.*

Input Data

In general, flow fields are based on non-regular two-dimensional and three-dimensional meshes. There may also be, as in finite element methods, nodal topological arrays, defining how the nodes are interconnected.

The data from the numerical simulation consists, at the fundamental level, of velocity, temperature and pressure. More complex systems will output stream and vorticity functions, tensor fields or many abstract variables such as are produced by κ-ε turbulence models.

Visualization

The usual techniques for visualizing CFD solutions are contours, vector fields, and isosurfaces in two and three dimensions. Additionally, CFD requires the visualization of streamlines and/or vorticity functions which are derived from the flow velocity and associated gradient fields. Often many scalars are combined on a grid point and so multi-dimensional techniques must be used.

Interactive CFD visualization is important in the tracking or steering of these solutions, although this may be restricted because of computational limitations. An important feature is the ability to select areas for enhancement or discarding and the specification of start points for particle tracks. Multi-grid solution techniques may require visualization of one dataset before the grid is refined: an interactive tool for this would be an important step forward in CFD visualization.

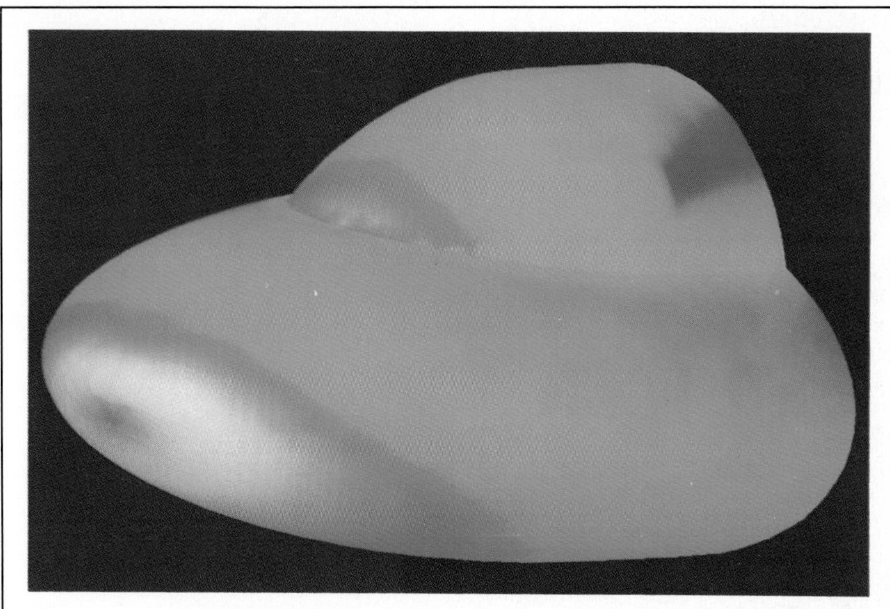

Figure 50. Pressure distribution on the surface of the Hotol forebody: The computed distribution of pressure on the surface of a double ellipsoid in viscous flow is shown for a free stream Mach number of 8.15 and at 30 degrees angle of attack *Visualization courtesy of K. Morgan and J. Peraire (Dept. of Civil Engineering, University College, Swansea) and M Vahdati (Dept. of Aeronautics, Imperial College, London).*

Figure 50 shows the viscous flow over twin ellipsoidal surfaces, analysing the performance of the Hotol forebody.

There are several products specifically for CFD visualization; some of the best known are:

Plot3D Plots flow data

GAS Graphical Animation System

RIP Remote Interactive Particle tracker

All are available from NASA Ames Research Centre, Moffat Field, California.

References

[Edwards89]

Edwards D.E., "3 dimensional Visualisation of Fluid dynamics", *AIAA Paper 89-0136*, **(1989)**.

[Kroos85]

Kroos K.A., "Computer Graphics Techniques for Three-Dimensional Flow Visualization", *Frontiers in Computer Graphics*, ed. Kunii T.L., **(Springer-Verlag, New York, 1985)**.

The two papers above have good coverage of the techniques used in CFD visualization.

[Bancroft89]

Bancroft G., Plessel T., Merrit F., Watson V., "Tools for 3D visualization in computational aerodynamics at NASA Ames Research Center", *SPIE* vol 1083, **(1989)**.

[Belie85]

Belie R.G., "Flow visualization in the Space Shuttle Main Engine", *Journal of Mechanical Engineering* vol 107, **(1985)**, *pp 27-33*.

These papers show the techniques in engineering use.

[Helman89]

Helman J., Hesselink L., "Automated Analysis of Fluid Flow Topology", *Three-Dimensional Visualisation and Display Technologies, SPIE* vol 1083, **(1989)**, *pp 144-152*.

[Helman90]

Helman J., Hesselink L., "Representation and Display of Vector Field Topology in Fluid Flow Data Sets", *Visualization in Scientific Computing*, ed. Nielson G.M., Shriver B., Rosenblum L.J., **(IEEE Computer Society Press, 1990)**, *pp 61-73*.

The two papers by Helman are also of interest.

6.17 Dynamics of Systems

Many dynamic systems exhibit non-linear behaviour. In the past solving systems exactly has been considered prohibitively expensive and engineers have relied on producing linear approximations to the real solutions and tried to avoid situations where these approximations would not hold. Recently, however, interest in non-linear systems has increased as it has been suggested that some catastrophic failures can be attributed to our inability to consider such non-linear behaviour in a design. For example, ship stability standards assume that the hull is stationary, because modelling the dynamic situation cannot be done with linear equations and so is considered too complex. Clearly it would be desirable to be able assure stability under more realistic conditions. Visualization is vital to such studies, as symbolic answers cannot be obtained in most cases.

Input Data

The state of the simulated system at a particular time is characterised by the values of a number of state variables. For the system to be studied geometrically, the system state is thought of as being represented by a single point in a multi-dimensional phase space, with one dimension for each state variable. As the system evolves, it traces a path in this phase space. For example, one represents the state of a pendulum as a point in a two-dimensional space, whose axes are angular position and angular velocity. Frictionless motion traces an ellipse in this space.

A typical system studied will have a small number of dimensions, perhaps five. A vector field is computed in this space and integrated at relatively high resolution to show possible paths of the system. Properties such as stability exponents, identities of attractors and other physical quantities are associated with points along these paths.

Visualization

The aim of global analysis is to uncover and illustrate the topological properties of the paths in phase space. Figure 51 on page 170 shows a phase portrait of two coupled pendula, coloured according attributes of the path. From the primary path data, various features are derived:

Invariant manifolds
: Surfaces on which paths are continuous. These divide the phase space.

Basins of attraction
: Sets of trajectories which converge on the same attractor.

Poincaré sections
: Cross-sections through phase space showing points where trajectories pass through the surface in a particular sense.

Figure 51. A phase portrait of two coupled pendula: The motion of the system is shown by the path in the three spatial dimensions, together with the two surface colorations. *Data courtesy of Dr T.Mullin, Oxford University. Visualization by the IBM UK Scientific Centre.*

For low-dimensional systems (energy conserving systems of two degrees of freedom or dissipative systems of one degree of freedom):

- single trajectories are shown as curves in 3D;
- invariant manifolds are surfaces in 3D, and are shown either as solid surfaces, or as sets of ribbons, both to suggest the direction of the trajectories and to make visible the underlying structures;
- basins of attraction are solids in 3D or sections in 2D;
- strengths of attractors are typically shown as equipotential surface in 3D or contours in 2D;
- Poincaré sections from many paths are scatter plots, showing the density of paths in given regions.

For higher-dimensional systems trajectories are presently projected to lower numbers of dimensions for display. As interest increases in these more complex systems, there is an urgent need to develop more interactive tools to help in exploring and understanding the data.

References

[Abraham84]

Abraham R.H., Shaw C.D., "Dynamics. The Geometry of behavior", **(Aerial Press Inc., 1984)**.

An excellent pictorial explanatory text on geometric dynamics.

[CompPhys90]

"Special Issue on Chaos", *Computer in Physics* vol 4 (5), **(1990)**.

This special issue on the subject of chaos contains a number of informative examples of the application of the techniques.

6.18 Program Visualization

There is almost unanimous agreement over the value of graphical programming, especially to end-users of CAD, presentation graphics and database systems. The benefits of program visualization are not as widely appreciated partly because research on visual programming has not as yet resulted in the creation of a well-defined environment for this purpose. The example described below was consequently based on ad-hoc graphics programming but it illustrates the value of visualization in the analysis of algorithms.

Line simplification is an essential task in cartographic generalisation. Algorithms for automatic generalisation of lines are still incapable of emulating the cartographer's skill. As part of an evaluation of line simplification algorithms the Douglas-Peucker algorithm was studied. Within cartography, it is widely believed that this algorithm is mathematically and perceptually superior to its competitors.

Input Data

The input consists of polylines of geographic features, such as coastlines. The visual analysis of the algorithm was simplified by tagging each point on the polyline with a tag value which corresponds to the significance of the point as calculated by the algorithm.

Visualization

Previous evaluations of the algorithm compared visually and numerically the output of the algorithm both with its input and with the output of competing algorithms. The concern was that the algorithm did not produce sensible results in many situations. The study therefore went beyond numerical comparisons and eye-balling of displays (both of which proved to be misleading and unreliable) and experimented visually with hypotheses about the properties of the algorithm and its behaviour under different data conditions to assess the validity of popular assumptions underpinning its use. The different lines of investigation utilised various visualizations of the computational process and of the data on input, output and tag values. This was initially undertaken as part of a brain storming exercise which revealed some conditions under which the behaviour of the algorithm contradicted expectations. Subsequently, a number of essentially two-dimensional displays were created and cross-referenced with the help of tabulated information to study a-priori hypotheses and conjectures about the method. Program visualization also included animated displays of the processing of specific polylines by the algorithm to assess the conditions under which the algorithm selects perceptually unimportant points and rejects perceptually more significant points. Visual feedback stimulated a chain of incremental and lateral thinking which led to the exposition of the inherent weakness's of the method and errors in popular assumptions about the process.

It is important to note that in this type of application it is often not possible to predict in advance the number or type of displays required let alone the manner in which they will be cross-referenced since they are problem specific and must mirror the researcher's diverging trends of exploratory thought. However, we believe that it is possible to construct generalised models and application-specific visualization systems which offer the required flexibility. This awaits further research since the currently popular model of scientific visualization does not cater for such divergent and lateral thinking.

References

[Myers90]

Myers B.A., "Taxonomies of visual programming and program visualization", *Journal of Visual Languages and Computing* vol 1 (1), **(1990)**, *pp 97-123.*

Reviews visual programming and program visualization.

[Visvalingam90]

"The Douglas-Peucker algorithm for line simplification : re-evaluation through visualization", ed. Visvalingam M., J.D.Whyatt, *Computer Graphics Forum* vol 9 (3), **(1990)**, *pp 213-228.*

Describes the use of visualization to study a particular algorithm.

6.19 Conclusions

The common characteristic of the applications listed is their concern with understanding the relationship between several variables in two or more dimensions. This varies from detailed study of the relationship of two variables over two dimensions, as in study of statistical indicators, to the modelling of the composition of fluids in three spatial dimensions and time.

The methods have all evolved in the context of specific applications and they appear to differ considerably from one area to another, most appearing to have characteristics special to the application. They also differ in their use of computer methods. Some, such as protein chemistry, have made heavy use of three-dimensional computer graphics and have well developed, fairly special purpose, solutions. Others, such as meteorology, while making heavy use of computers for simulation have well established two-dimensional graphics methods. Other areas, such as the study of dynamics, are only now becoming practicable with the development of computer visualization methods.

If the benefits are to be brought to a wider audience, more general systems have to be used. Such systems are now coming into existence and some experimental trials are being reported, particularly in the earth sciences, but none of these general purpose visualization tools yet appear to be in regular use in the reported applications. In consequence it is too early to give an evaluation of them.

Much of the present emphasis in scientific visualization is on the display of data in three dimensions, and three dimensions together with time. This is probably because of the immediate appeal of the results. However, there are some applications, for example in cartography, which have smaller numbers of dimensions but which are not well catered for. For instance, one cannot easily investigate alternative methods of presentation of the same data. Very flexible systems are needed to make this easy. Similarly, less work has been done on the difficult problem of the display of high-dimensional parametric data.

Progress in these areas will help considerably in our understanding of some complex and important problems.

6.20 Key References

See each of the sections on particular applications for their key references.

Chapter 7

PRODUCTS

Edited by Julian Gallop

7.1 Introduction

Previous chapters have presented a number of aspects of scientific visualization. The purpose of this chapter is to provide a brief survey of current examples of visualization products.

Although definitions have been provided elsewhere it is worth restating that a visualization system is an integrated whole, supporting the effective exploration of complex data by visual means. It is difficult to meet this goal at the present time and what we present here are systems which are attempts in that direction. For example, although a library of graphics software does not adequately allow a user an easy way of selecting a visualization technique, some library based systems are included here. Again, ideally a visualization system accepts multidimensional data, but at present we have to relax this requirement for inclusion.

Although this chapter focusses on visualization software products as such, it is important to also understand the role of technologies that underly visualization. Some of the most important are described in appendix "Enabling Technologies" on page 217:

- hardware
    - hardware platforms, including superworkstations
    - graphics input
    - graphics output
- graphics software
    - 2D graphics software
    - 3D graphics software
    - window systems
- user interface toolkits
- image tools
- database management systems

The examples in this chapter are restricted to products that can be obtained for one's own use.

To our knowledge, none of the visualization products mentioned here are suitable for data from fields such as computer science, where structures such as networks and graphs are common, including examples such as program structure and execution profiles. Although exploring such structures by visual means is visualization, it was not a primary consideration at the workshop and different techniques and skills are likely to be needed.

The chapter is based on [Haswell90], updated where necessary. In many cases, the information is from brochures and the literature. Where possible this is backed up by practical experience.

7.2 Visualization Software Categories

Visualization software has evolved over a period of time and three categories can be identified which tend to mirror their evolution. The categories are not rigid. In general the older the category the less power, memory and storage required to run them, which makes the software in the first category suitable for use in PC or terminal/mainframe environments and the most recent developments only suitable for the most modern, powerful supercomputers or superworkstations. At the same time, because the first and second categories have been around longer, more applications have been developed using them and many products in the market fall into these classes. It is likely that in the future, tools using the most modern techniques will appear but at the present these techniques are very much in the experimental and development stage and will need some time to mature.

7.2.1 Graphics Libraries and Presentation Packages

This is the traditional method for creating new ways to view and analyse data. The libraries interface directly to graphics hardware or provide graphics functionality in software. The user has to supply nearly all the components to support the application: the main program, the user interface, data handling and geometry mapping. The most basic libraries only supply an interface to the graphics devices

(PLOT10, HCBS) and some higher level libraries handle more sophisticated graphic entities such as axes, curve drawing and so on (DISSPLA (Precision Visuals), GL (Silicon Graphics), the NAG Graphics Library and Doré (Stardent)).

Many PC based packages such as Harvard Graphics, Slidewrite and CricketGraph have taken on board the user interface functions to provide more user-friendly software, but still require a great deal of user effort to achieve good results.

The advantage of this type of software is its flexibility and direct control but it suffers from the disadvantage of the large amount of time invested in writing and supporting code.

7.2.2 *Turnkey Visualization Systems*

For the systems in this category, the user provides the data and instructions to the main program. The system supplies the main program and rendering and usually has an attractive user interface.

The user does not have to program these packages and can obtain results very quickly. However in general the packages cannot be modified or extended and therefore often only provide a part of the solution a user requires, hence the use of the term "turnkey" here - (which is not intended to imply that the system comes with fixed hardware).

Some of these products provide additional flexibility by means of a procedure library (for example, PAW, SunVision and UNIRAS), but do not allow modules in the interactive turnkey product to be substituted.

This category can be subdivided further into systems which primarily handle geometric data (Geometry Viewers) and those which handle property data, usually in association with geometric data also (Turnkey Systems for Scientific Data) - geometric and property data are discussed in the chapter on Data Facilities (see "Geometric and Property Data" on page 90).

7.2.2.1 *Geometry Viewers*

Soon after the term ViSC came into common use, the name "Visualization Software" typically referred, in the superworkstation market, to an interactive geometry viewer for the display and animation of 3D objects. Geometric data could be accepted, but usually not property data. The objects could take the form of polygons, meshes or NURBS and could have colour associated on a per object, per facet or per vertex basis similar to PHIGS PLUS (see appendix "3D Graphics Systems" on page 224). Either the input format of the data could be fixed or a mechanism would be provided for creating data filters to convert to an internal format. Usually these packages claimed to provide editing facilities but in fact only allowed limited changes to the data, such as: the overall colour of an object (including ambient, diffuse and specular reflection properties); light properties (colour, whether directional or point etc.), typically providing up to eight distinct light sources; shading techniques - such as wire, flat or Gouraud; camera position (or the viewpoint from which to view the object); interactive 3D transformations of an object, camera or light source (only one item at a time).

7.2.2.2 Turnkey Systems for Scientific Data

These systems allow a user to supply property data, possibly in association with geometry data, and allow a user to select from a fixed set of operations on that data.

Many products in this category are application-specific and examples in oil exploration, molecular modelling and architectural modelling are common but of limited use in other fields. They are also very often only available on a very few hardware platforms in common use in those industries for commercial reasons.

More general examples are the UNIRAS interactives, PV-WAVE (Precision Visuals), Data Visualizer (Wavefront), SunVision interactive programs (SUN) and VoxelView (Vital Images).

They have all reached a high level of maturity and many users applying visualization to their work are probably using one of these packages.

7.2.3 Application Builders

These offer a series of modules linked by interfaces which are connected interactively at runtime. They combine features from the previous categories by providing "turnkey" solutions for individual parts of the program and the flexibility to customise the final solution adopted. The supplied modules can be replaced by user-written modules as required, providing they conform to the data input/output interface conventions. Some limitations exist; for example, in AVS the viewers are essentially monolithic - the network editor provides the route to build applications.

In these systems virtually everything the user needs is provided by the program. The user has only to direct the path of the program and the data. In some systems, the user can insert application computation modules which would allow the application to be steered by the user through the medium of the visualization system, a powerful capability.

Applications are generally constructed by a mouse driven interface, manipulating icons on the screens and linking them with data paths. Once the required modules have been connected and built the program can be executed. New applications can be prototyped very quickly by connecting modules in different ways but the user needs to know how to manage the flow of data through the network and how to extend the module set.

Examples are AVS3 (Stardent), apE (Ohio Supercomputer Centre) and Khoros (University of New Mexico). More advanced application builders are currently under development and new advances in visualization techniques and software will extend and improve the application builder functionality.

Although the visualization products mentioned here can be used to display data from a wide variety of fields, the gap between what the scientist would like to input and what is accepted by an application builder product is often wide. As an example it is not possible to display mathematical equations directly with these packages, and yet suitable application-dependent software exists in many cases. At the present time these products are not mature enough to have had substantial numbers of packages built around them, but perhaps with time more application-specific tools, following the example of the viewers from AVS, which are built on top of the general-purpose software will become available.

7.2.4 Categories explored in this Chapter

The remainder of this chapter will concentrate on software products falling into the second and third categories. This does not imply that good visualization work cannot be done with software libraries and PC packages but reflects the good understanding and wide experience of these systems that already exists.

Although PCs are abundantly available in the UK academic community, the workshop was able to regard graphics software on PCs as addressed by an existing recent publication [AGOCG90].

Additionally it is probable that restrictions on internal data transfer bandwidth and graphics display facilities would limit the usefulness of the current PC products as hardware platforms for the majority of software products in the turnkey and application builder categories. PCs can however be utilised as X servers for such software running on superworkstations, supercomputers or mainframes and connected via X25 or Ethernet. Users of PCs therefore can have access to visualization systems remotely and view the results at their desks, albeit at lower graphics resolution and with a time penalty introduced by current network performance.

7.3 Examples of Software Products

This section presents examples of visualization software, based on a study of the literature, some conversations with suppliers and in one or two cases some actual use. The section is divided into the categories just defined.

7.3.1 Factors

A number of factors can be used to describe visualization systems. For each system, some are presented in a table, others are discussed in the text.

- availability - this breaks down into a number of subheadings:
  - suppliers and software support
  - example configurations on which it is available
  - source availability

  It would in future be useful to provide information on dependence which would indicate how easy it would be to port the visualization system to other hosts. Dependencies would include programming language, user interface toolkit and system dependencies (for example, inter-process model assumed), as well as the underlying graphics software.

- cost - rather than give precise cost figures which constantly change, cost categories are provided.

  no cost - some visualization products are available at no cost from generally academic sources. Products from academia generally come with source-code and hence enable tailoring to specific needs or porting to new platforms.

  low cost - it is not uncommon for academic establishments having designed a substantial software product, rather than to stretch their limited resources, to charge a nominal fee to cover distribution costs (for example apE).

  bundled - some products are available at no cost bundled with products from hardware vendors. The software from vendors is mostly hardware-dependent (with AVS increasingly becoming an exception to this) and available in object-code only. By contrast with the products free from academic sources, the bundled products would normally be supported by the hardware vendor.

  commercial rates - products that cost real money.

- procedural library - some of the turnkey systems have procedural libraries, allowing an application program to invoke the visualization facilities. For an application builder, this is assumed as given, so is not presented explicitly in the table.

- application steering - some application builders allow application modules to be included within the framework of the visualization system, enabling the application to be steered.

- functional distribution - in some of the visualization systems, it is possible to distribute the functions simultaneously on multiple hardware platforms via a network, with the visualization system conveniently packaging the system calls necessary to achieve this.

- underlying graphics system. For many users this should be invisible. It becomes important if the user is reconfiguring the system to add display methods, as the graphics system may well need to be called by the user in such cases. It is also important when choosing a system. The underlying graphics system can crucially determine the speed of operation. If the underlying graphics system is 2D, the system will be unable to take advantage of 3D graphics firmware.

- support for X11 protocol, allowing an X server to be used.

- data model - the main classes of data accepted by the system. This can be geometry data or property data, which can be images, volumes or more generally multidimensional data.
- import - the specific data format imported by the visualization system.
- export - the specific data format exported by the visualization system. This includes formats for exporting pictures for viewing or for hardcopy; possibilities could include CGM [ISO(8632)87], RIB (RenderMan Interface Bytestream) [Upstill90] or PostScript [Adobe85].
- visualization techniques.
- the quality and ease of use of the user interface.
- references to more information about the product in the literature.

7.3.2 Examples of Turnkey Visualization Systems

7.3.2.1 Examples of Geometry Viewers

Examples of the early geometry viewers (generally these could be obtained bundled) were:

- **Personal Visualizer** bundled on Silicon Graphics systems (originally from Wavefront, from whom there is now an enhanced version);
- **Visedge** from Alliant.

In addition, version 1 of Stardent's AVS was of this sort - its functionality is now in the Geometry Viewer of current AVS releases.

Visedge and Stardent's Geometry viewer are closely based on PHIGS PLUS capabilities whilst Personal Visualizer is based on GL, the proprietary graphics library from Silicon Graphics.

These examples only allow the manipulation of one object at a time and so it is not possible to have some items following one animation path whilst others follow a different path (such as the rotating wheels of a car while the whole car moves forward).

There is no method here for redefining how the data is to be displayed (such as swapping the z-value with the colour index) - a new data file would need to be generated with the new format.

Data filters providing interfaces to commonly used software packages are often provided, for example to one or more of the following CAD packages: MSC/NASTRAN, PATRAN, SDRC's I-DEAS. Visedge and the Geometry Viewer allow the user to import data from a variety of sources by allowing data filters to be added.

An advantage of Visedge and the AVS Geometry Viewer is that they avoid heavy use of pop-up menus, for example to change the surface colour of an object, where they use a slider bar method to interactively change the colour.

Personal Visualizer may receive data, interactively generated by a simple 3D modelling package called QuickModel (bundled free with the higher performance ranges from Silicon Graphics), along with QuickPaint (which can be used to generate texture maps).

7.3.2.2 Examples of Turnkey Systems for Scientific Data

Data Visualizer

Data Visualizer is an interactive visualization system from Wavefront. Data input is mainly mesh data (either vector or scalar) and the input formats include Plot3D and NetCDF. Mesh formats include regular grids, regular grids with irregular spacing, irregular grids and unstructured grids.

Facilities include: data probe, cutting planes, particle trace, volume rendering and colourmap editor. In its most recent release, multiple datasets can be presented using the product's view manager. 2D graphics can be used, not just 3D. In a future release it is expected that some aspects of the application builder category will be provided.

Data Visualizer can be used with other Wavefront packages such as Personal Visualizer and the Advanced Visualizer.

supplier	Wavefront Technologies
s/w support	yes
source code supplied	no (except that sufficient source to add a data reader is supplied)
cost	commercial rates
example configurations	SGI 4D systems, DECstation 5000 PXG Turbo, IBM RS6000, HP Apollo 700; 24 bit colour is necessary
procedural library	no
functional distribution	limited - data readers can be on another system
underlying graphics	SGI GL, PEX/PHIGS on DEC systems
X support	no
data model	vector and scalar data sets on a mesh
import	PLOT3D, NetCDF, WAVE - readers can be added by users
export	Colour PostScript, WAVE, TIFF

Table 1. Data Visualizer Summary

Intelligent Light Products

Intelligent Light provide a range of products suitable for visualizing data: CADimator, IVIEW and 3DV. CADimator is aimed at the display of CAD/CAE data; IVIEW for scientific/engineering data; and 3DV for more general data. An animator, which includes interactive and scripted definition of an animation sequence is provided. An additional tool VIDEOTOOLS imports images from several external sources, composes new images by combining several and outputs to videotape. FieldView is a relatively new tool for the display of scalar and vector datasets and the display techniques include volume rendering.

The five products, which are compatible, have been implemented on top of Doré and so can be relatively easily ported to new hardware platforms after the implementation of Doré.

supplier	Intelligent Light
s/w support	yes
source code supplied	no
cost	commercial rates
example configurations	Silicon Graphics, HP Apollo, Stardent
procedural library	toolkit available
functional distribution	no
underlying graphics	native graphics libraries, also Doré

Table 2. Intelligent Light Product Summary

MPGS

MPGS (Multi-Purpose Graphics System) [Grimsrud89] from Cray provides a distributed approach to graphics, by allowing a user access to the computational power of the Cray and to the graphical power of a Silicon Graphics 4D workstation. Programs are simultaneously executed on both systems to provide the data connection and, to reduce network traffic, the data transfer is limited to visible parts of the data in the required scene. Another technique used to limit data transfer is, where appropriate, to pass high level commands to the graphics workstation for processing; interactive graphics such as the transformation of objects is handled this way.

supplier	Cray
s/w support	yes
source code supplied	no
cost	commercial rates
example configurations	Cray + Silicon Graphics
procedural library	
functional distribution	yes
underlying graphics	Silicon Graphics GL

Table 3. MPGS summary

The emphasis with this system is on computational modelling type applications as it provides the following display methods: vectors displayed as arrows, particle tracing, contouring and isosurface representations. It can handle finite difference, finite volume and finite element data where the data is assumed to be unstructured.

NCSA Tools

The National Center for Supercomputing Applications (NCSA) at the University of Illinois at Urbana-Champaign obtained a grant and equipment from Apple Computers in order to provide visualization tools for a Macintosh II. Some of the early versions of the resulting software are freely available and require a Macintosh II with 8 colour planes. The tools include programs to display and analyse surface data (as contours), image and volume data; and create animations. Commercial products are being marketed for the Macintosh II by a new company called Spyglass.

More importantly for the workstation market, are the X11 implementations that followed the Macintosh tools (both versions of the freely available software are available via anonymous ftp). The X11 tools currently provide tools for the display and analysis of image data, with XImage [XImage89], and for volumetric data with XDataSlice [XDS89]. XImage (along with the Macintosh counterpart) allows the user to

- display data values in a spreadsheet format,
- use cartesian or polar coordinate systems,
- display the data as black and white contour or colour shaded plots,
- display a coloured frequency histogram for the data values,
- generate animation sequences of the images.

XDataSlice provides facilities for displaying data in spreadsheet format, arbitrary slicing and dicing of volumetric data, surface tiling, interpolation of images to generate a more complete volume, and saving frames for an animation sequence. An additional tool, for use in presentation quality images, is called CompositeTool allows the user to annotate images and include representations of the colour palette used.

A PC product from NCSA, PC Show, displays data files as colour images on an IBM PC compatible for EGA, VGA and other graphics cards. Multiple images can be animated and colour palettes can be manipulated.

An integral part of the NCSA tools is the self describing data file format, called "Hierarchical Data Format" (HDF) (see "Data Formats" on page 102).

The product information given below is for the version obtainable via ftp, not the Spyglass Macintosh version.

supplier	NCSA via anonymous ftp 128.174.20.50
s/w support	new versions are produced
source code supplied	yes
cost	none
example configurations	Apple Macintosh II with minimum 8 planes, Unix workstations, Cray
procedural library	a procedural interface to HDF is provided
functional distribution	no
underlying graphics	X11 on Unix workstations
X11 support	yes
data model	images, multi-dimensional data
import	HDF
export	HDF

Table 4. NCSA Tools summary

OASIS

OASIS is a solid modelling system from Cray originally developed by Gray Lorig at Rensselaer Polytechnic Institute (RPI). It has a small number of very powerful primitives, including equi-value 3D surfaces and superquadrics. Its only rendering techniques are wireframe and ray trace (nothing in between). It uses object-oriented programming techniques to specify animation of any parameter of any object, allowing fully general choreography of any scene. It runs on the Cray, the Cray and Silicon Graphics combined and the Silicon Graphics as hosts, has input filters for a number of geometry systems such as MOVIE-BYU and output to many devices including, at RAL, CGM and video.

PAW Program Suite

PAW, the Physics Analysis Workstation from CERN ([Brun89], [Goosens89] and [Vandoni89]), is an interactive data manipulation and analysis program suite, including statistics and minimization, and includes many display facilities. Some 3D displays are possible now; others are being written, for example the support of additional types of display output such as spheres. There is a defined graphics interface, which can be implemented on 2D graphics packages, including the 2D graphics standard GKS: the graphics package per se is not supported as part of the PAW system. PAW can handle errors provided by the user with the data. The user interface is by text commands or is menu driven. The macro language is extensible by the user and there is comprehensive on-line help. Hardcopy output from PAW depends on the underlying graphics package, but typically includes both CGM and PostScript files.

supplier	CERN
s/w support	yes
source code supplied	no
cost	no cost to Physics Depts in academic or other cooperating institutions
example configurations	Sun, HP Apollo, non-workstations up to IBM 3090; can run Tektronix 4010 emulators, including PC/Emutek
procedural library	yes
functional distribution	on VAX/VMS, IBM or Unix systems
underlying graphics	2D graphics, including GKS
X11 support	yes
data model	2D; others are programmable
import	any, via programming interface
export	depending on graphics interface: CGM, PostScript

Table 5. PAW Program Suite summary

PV-WAVE

PV-WAVE [Charalamides90], from Precision Visuals, provides an interpreted command interface along with some macros to enable the user to navigate through large datasets. The interface resembles a typical programming language with constructs such as "if ... then ... else" and "for" and "while" loops which can be combined to form subroutines (macros); this is based on the Interactive Data Language (IDL). Routines are included to assist the user to read in multi-dimensional data (of various formats) and to build prompted or menu-based user interfaces. Facilities used to display the data include: plotting, contouring, and surface generation (using meshes). PV-WAVE appears to be strong at the display and analysis of image data. For multi-dimensional data the user can choose (possibly by named options from a pop-up menu) which elements are to be used and with which display technique, while several windows could be used to display different aspects of the data.

supplier	Precision Visuals
s/w support	yes
source code supplied	no
cost	commercial rates
example configurations	from PC + Emutek to SGI workstation
procedural library	a module-like interface is provided
functional distribution	no
underlying graphics	2D graphics system; an X interface is provided
X11 support	yes
data model	images, multi-dimensional data can be input (can be sliced using progamming interface)
import	programmable
export	PostScript

Table 6. PV-WAVE summary

The output from PV-WAVE can be produced as a PostScript file. Having generated a series of commands to process data in a required way these commands can then be "compiled" into an application rather than interpreted. There appears to be no interaction with the data once it been displayed, for example to interactively change thresholding of the data or to vary the image quality. The underlying graphics library is 2D which prevents interactive manipulation of 3D objects although, since a mechanism is provided to record and play back animation sequences, non-interactive animations can be generated. PV-WAVE is currently available for DEC and Sun workstations.

More recently a point and click interface has become available. Also announced and shortly to be made available is an interface to the NAG numerical library, PV-WAVE:NAG.

SunVision

This is a product for the Sun SparcStation range [Sun90]. It provides both interactive systems and toolkits for the display and analysis of: objects similar to those in PHIGS PLUS, 2D images and volumetric data.

A procedural library is available allowing most of the functions in the interactive programs to be accessed from an application program - the volume rendering facilities are an exception to this.

There are three methods for the display of volume data:

1. **cube mode**: displays the volume as a cube, with the voxel values texture-mapped onto the surface and, by interactively slicing away parts of the cube, features within the volume can be displayed;

2. **lightbox mode**: treats the data as a sequential series of 2D slices and allows the user to pan through the volume one slice at a time;

3. **cloud mode**: displays the surface by clouds of points; the number of points can be increased or decreased to obtain the required image quality and performance.

A geometry renderer provides most of the functionality of RenderMan for high quality rendering.

SunVision also performs image processing and animation.

supplier	Sun Microsystems
s/w support	yes
source code supplied	no - except sufficient source to write application programs
cost	commercial rates
example configurations	from Sun SparcStation with 8 bit graphics to Sun SparcStation with (announced) VX and MVX graphics accelerators
procedural library	yes
functional distribution	no
underlying graphics	Sun's own XGL
X11 support	no
data model	geometry, images, volumes
import	Sun vff, AUTOCAD .dxf files and others
export	Sun vff

Table 7. SunVision summary

UNIRAS

UNIRAS [Jern89] provides both libraries and interactive systems for the display of a wide range of data. Presentation of scientific data can be achieved via the interactive program UNIGRAPH.

Via its libraries it provides many of the facilities in PV-WAVE but in addition is able to display: cartographic projections, simple volume data (the data can be represented as coloured contours over blocks, where groups of blocks can be removed in order to view within the volume), and PHIGS PLUS like objects. Although it does allow the display of 3D data this is achieved by first converting to 2D and so 3D interactions such as rotating objects are not possible.

UNIRAS is available on a wide range of hardware platforms and attempts to provide everything possible to display static images on systems with very poor graphics facilities, including hard-copy devices, through to Sun workstations (a Sun X11 implementation now exists which may extend availability to most Unix workstations).

A user interface management system USEIT is available, to assist with the development of UNIRAS applications.

The UNIRAS package is available at commercial rates, though a subset of the tools can be purchased at a reduced rate. The UK academic community have a special deal for the UNIRAS sites which allows sites to purchase the software for the whole site on a particular machine range for a single annual charge.

supplier	UNIRAS
s/w support	yes
source code supplied	no
cost	commercial rates (there exist special arrangements for academic institutions in the UK)
example configurations	from PC + Emutek to SGI 4D workstations (also available on Cray)
procedural library	yes
functional distribution	no
underlying graphics	2D graphics, X11
X11 support	yes
data model	geometry, images, volume (some slicing capability)
import	raw ASCII data
export	CGM, PostScript, UNIPICT

Table 8. UNIRAS Interactive Programs summary

VoxelLab/VoxelView

A volume rendering package called VoxelLab is bundled free with the higher performance ranges from Silicon Graphics. It is based on VoxelView [VoxelView89] a commercial product from Vital Images.

VoxelView's volume rendering methods take advantage of Iris workstation graphics chips and some of the internal loops are implemented in microcode. A great deal of effort has been expended on advanced/optimized/enhanced volume rendering methods.

VoxelView relies heavily on hierarchical menus for a great deal of the interface, which can lead to several menu traversals to get to the required selection. More satisfactory aspects of the user interface are: the use of slider bars, where appropriate, for inputting data values, and a Macintosh-style icon-based system for selecting data files. For files containing volume data the icon shows a small rendering of that data, for readable directories a folder and a "no-entry" sign indicates a non-readable directory.

The data value of the voxel (ranging from 0 to 255) can be mapped to corresponding opacity values (also from 0 to 255), where the mapping can be defined by a function with minimum, maximum and curvature values. With this technique it is possible to display, for example, high valued voxels more transparently than the low values and so enabling the user to view faint or obscured details. Other facilities are also available to enhance features, such as thresholding out voxels from an irrelevant subrange or the redistribution of voxel values from one range to another. Since volume rendering is very time consuming, VoxelLab and VoxelView allow the user to store a sequence of renderings which can then be played back in real time.

The information below applies to VoxelView.

supplier	Vital Images
s/w support	yes
source code supplied	no
cost	commercial rates
example configurations	Silicon Graphics 4D workstations
procedural library	no
functional distribution	no
underlying graphics	SGI graphics libraries
X11 support	no
data model	volumes

Table 9. VoxelView summary

WinVis90

WinVis90 [Collins90] is obtainable from the European Visualization Centre at the IBM UK Scientific Centre, free under a software evaluation agreement. The user agrees to provide written feedback after no longer than 12 months as to the use they have made of the code, its performance and deficiencies. The code runs on IBM RISC/6000 machines that have the 24 bit High Performance Graphics Adapter. WinVis90 allows interactive visualization of equations, 3D scalar fields (both regular grids and scattered data), display of images, colouring of geometries by scalar fields, data probes of various types for exploring scalar fields, batch command facility for demonstration purposes, image capture for video animation, rotation, translation and scaling of scalar field contours, the ability to have contours move on separate paths, performance tuning under user control, joint display of geometry and isosurfaces, display of molecular database pictures and a tailorable user interface.

supplier	IBM Winchester, UK under evaluation agreement
s/w support	some
source code supplied	no
cost	none
example configurations	IBM RS6000 systems with 24 bit graphics from model 320 to model 730
procedural library	no - planned
functional distribution	no
underlying graphics	PHIGS or SGI GL
X11 support	no
data model	geometry, images, multi-dimensional data, mathematical functions
import	fixed, but originators can extend
export	fixed, but originators can extend

Table 10. WinVis90 summary

7.3.2.3 *Other Turnkey Systems*

A useful class of turnkey system that is worth mentioning but is not considered in detail here, are systems which are primarily mathematical, but also provide display capabilities. These include:

- Derive (Soft Warehouse Inc),
- JMP (SAS Institute Inc),
- Maple (University of Waterloo),
- MathCAD (Adept Scientific Micro Systems Ltd),
- Mathematica (Wolfram Research Inc),
- Matlab (Mathworks),
- Resolution (P.S.Squared Ltd),
- S-PLUS (STATSCI),
- STATGRAPHICS (STSC Inc).

Other systems already considered can be used in this context. For example, WinVis90 allows interactive visualization of equations and AVS can accept data from Mathematica and display it.

7.3.3 *Examples of Application Builders*

7.3.3.1 *apE 2.0*

The apE system (this acronym was initially for "animation production Environment") uses data flow diagrams and uses a visual programming technique to allow the users to build applications - see [Anderson89] and [Dyer90]. This software is network transparent and allows the user to incorporate routines that execute on different systems. When presenting the diagram of routines and connections, apE includes a picture to represent the routine. These can be thought of as picture tiles, with the name written on the back. The user can flip over a single tile to see the name or ask to be shown by default the name for all tiles rather than the graphic.

This software has been written in C and rather than being tied to one graphics library, a common interface was designed, a high level layer was written to the common interface and the underlying routines were implemented as three separate ports to: SunView, Silicon Graphics' GL, and X11. A later version will probably be based on PHIGS. Tools are provided for the display of: 2D and contour images; 3D isosurface, volumetric and terrain images; polygonal data. Animation is possible, with some user control over the speed of playing back animation sequences.

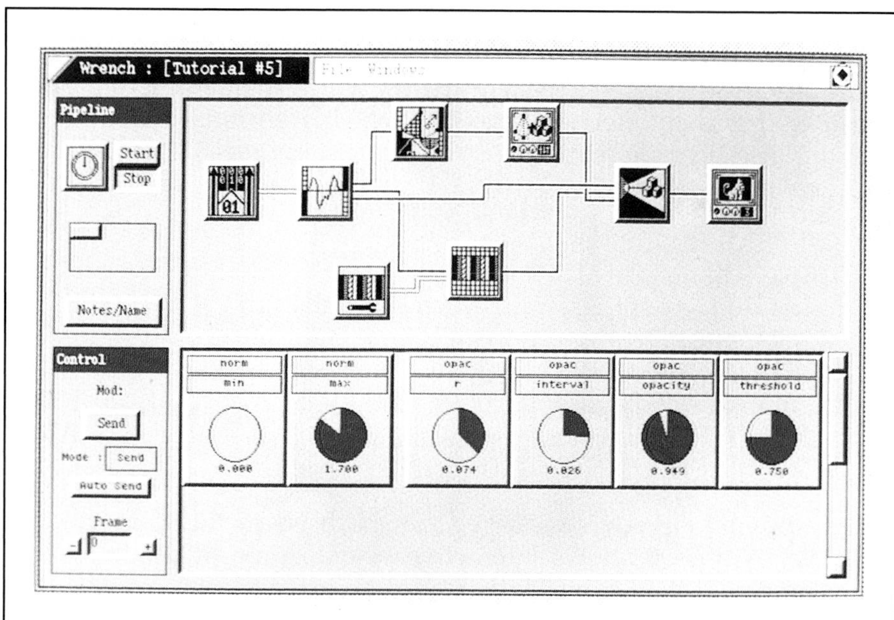

Figure 52. An example of an apE data flow diagram:

supplier	Ohio State University
s/w support	bugs can be reported
source code supplied	yes
cost	low cost
example configurations	Apple Macintosh II; Unix workstations, including Sun, Silicon Graphics and Stardent; Cray supercomputer
application steering	yes
functional distribution	yes
underlying graphics	X11 or SGI GL
X11 support	yes
data model	geometry, images, volumes
import	HDF, many image formats
export	many including PostScript

Table 11. apE summary

Additional tools are provided for image format conversion, image processing and compositing. The source code is distributed along with the binaries. An example of an apE screen, including a data flow diagram, is shown in Figure 52.

7.3.3.2 *AVS*

AVS (Application Visualization System) - described in [Upson89a] and [vandeWettering90] - derives from Stardent and is released on their workstations and increasingly on other suppliers' systems also. AVS2 is the currently released version of AVS with AVS3 having recently completed beta testing at the time of writing. From the AVS2 release onward, the software is divided into both interactive systems and toolkits for the display and processing of geometry, image and volume data.

There are four interactive systems in AVS2: the Geometry Viewer, the Image Viewer, the Volume Viewer and the Network Editor which allows application building.

Geometry Viewer - displays PHIGS PLUS-like objects and comes with example data filters. The Stardent 2000 version takes advantage of a hardware "fast" sphere useful in the display of molecular data and a hardware implementation of texture mapping.

Image Viewer - allows the user to read in an image, which is an X pixmap, then interactively apply some image processing techniques to it. With the data preprocessors it is possible to crop, downsize (sub-sample), mirror, transpose and interpolate the data values. Having displayed the image it is then possible to magnify up or down the image, but only by a set of predefined values and only about a central point. Apart from modifying the colourmap of an image it is possible, by using parameter controlled values, to clamp the pixel values to a chosen range, to modify the contrast and to perform a histogram stretch to equalise the distribution of values over a chosen range. Another method available with this viewer is the ability to generate 3D meshes by mapping the pixel values onto the z coordinates. The resulting mesh can be manipulated, by switching to the geometry viewer, and can also have the original image texture mapped over the surface (this could be useful in generating 3D terrains from 2D images). It is possible for the user to add functionality to the image viewer, by incorporating image processing networks, defined with the network editor.

Volume Viewer - in addition to the data preprocessors available with the image viewer this viewer provides several display techniques and rendering techniques. The display techniques include: orthogonal and arbitrary slicing, bubbles (where each data value is represented as a sphere), dot surface (having created an isosurface the surface is represented as a mesh of dots), and isosurface tiler (which uses the "Marching Cubes" algorithm [Lorensen87] to create an isosurface).

Network Editor - uses visual programming techniques to allow the user to build new applications from a set of building blocks called modules (which resemble task-dependent subroutines in an application). The network defines an application by showing the modules along with the connections that show the flow of data from one module to the next, in a form that resembles a flow diagram. To generate this network the user drags modules from a palette and is able to make and break lines between modules.

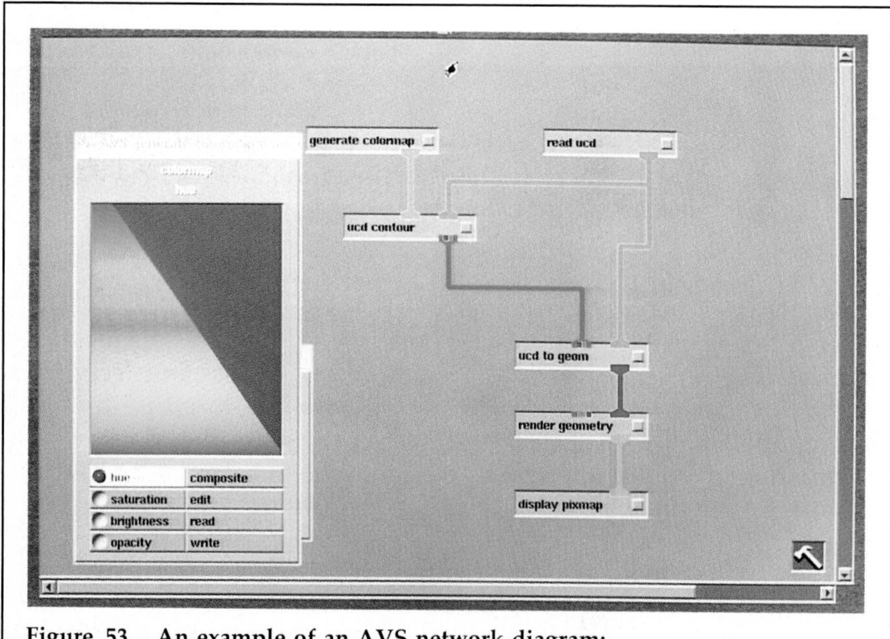

Figure 53. An example of an AVS network diagram:

Each module can transfer data via its input/output ports and a colour-coded band gives a visual indication of the data type. Only compatible ports can be linked together. The flow of data through the network is strictly acyclic but it is not unusual for applications to require some form of feedback. Text is used to label the modules in the palette and the network diagram. Numerous compiled modules are provided for the user to use but for tasks not covered, such as a particular image processing algorithm, the user can quite easily write or modify existing code, including a collection of user-contributed modules. An example of an AVS network diagram (and also a colour map window) is shown in Figure 53 on page 198.

One useful tool available as a module for new applications using colourmaps but also used in the image and volume viewers is an interactive colourmap editor.

With the interactive viewers and the user generated applications, it is the user's decision as to how many windows showing different views are shown at any one time. However, some modules are CPU hungry and may cause severe delays in response time with simultaneous views.

Apart from simple data types (byte, integer, real, string) more complex formats are available to the application developer: geometry, colourmaps, pixel maps, fields. Fields are the general n-dimensional data type that can map onto rectilinear and irregular coordinate systems enabling the representation of simple mesh data. The limitations with this type are that the data is all of the same simple data type (for example all real) and that they do not allow the representation of unstructured meshes.

Standard output formats include (monochrome) PostScript but it is possible for the user to generate modules for additional file formats.

Extensive on-line help is provided.

This software was originally written for the Stellar hardware (now Stardent 1000 and 2000) using a combination of PHIGS PLUS (at that time PHIGS+) with X11, along with limited use of XFDI (Stellar's 3D extensions to the X Window System) calls. After the merger of Stellar with Ardent, to form Stardent, the Ardent hardware required a new version of AVS to be implemented on top of the Doré graphics library. Subsequently additional vendors, including Convex, Evans & Sutherland and DEC have licenced the software for distribution with their products.

With AVS3 an additional viewer for the display of 2D and 3D graphs is provided. AVS from this release can, like apE, run in a distributed environment - modules can be compiled and executed on other systems and incorporated into networks. Although, at present there is no way of specifying which modules can run in parallel this seems like the next step from here. Data can be passed upstream in a network and networks can now be iterative. Data probes can return the coordinates at a selected point. Colour PostScript is now supported. New data types include the atom bond (in preparation for a chemistry viewer) and unstructured cell data (which allows the representation of unstructured mesh data - in preparation for a CFD viewer); some user control over data types also becomes possible. From this release it has been strongly suggested by Stardent that new viewers will be unbundled and made available at a price.

The following information applies to AVS3.

supplier	Stardent
s/w support	yes
source code supplied	not normally
cost	the situation regarding bundling and costs appears to vary between suppliers
example configurations	workstations including DEC 5000 Ultrix, Evans and Sutherland ESV series, IBM RS6000, Stardent; supercomputers including Cray, Convex and Floating Point Systems
application steering	yes
functional distribution	yes
underlying graphics	PHIGS+ or Doré
X11 support	yes
data model	geometry, images, volumes, unstructured cell data; also multi-dimensional defined
import	Mathematica, PDB, MOVIE.BYU, AVS fields; a filter to input HDF exists, although not at present supplied with the release
export	AVS fields, colour PostScript

Table 12. AVS summary

7.3.3.3 *Khoros*

Khoros includes some simple support for visual programming when building its data flow network and contains code generators for extending the visual language and adding new application packages to the system. There is an interactive user interface editor, an interactive image display package, an extensive library of image and signal processing routines, and 2D/3D plotting packages. Khoros started life as a purely image processing environment and this is reflected in the structure of the internal data format supported by Khoros, VIFF which contains images and multi-dimensional (banded) data.

It runs on X11R4.

supplier	University of New Mexico via ftp from (129.24.13.10); electronic mail contact is rasure@chama.unm.edu
s/w support	no (can report bugs)
source code supplied	yes
cost	none
example configurations	Unix workstations with X11R4 and PC/X11R4; Cray
application steering	some support
functional distribution	not available in this release
underlying graphics	X11
X11 support	yes
data model	images, multidimensional data; structured and unstructured data
import	VIFF, raw data in a number of forms, X bitmap files (xbm), BIG, Sun raster files, TGA, DEM, DLG and several specialized file formats
export	VIFF, Sun raster files, PostScript, BIG, xbm

Table 13. Khoros summary

Khoros, in the current release, does not support remote execution of modules; communication between modules in the Khoros network is realized through temporary files, rather than Unix pipes and sockets.

Descriptions of Khoros can be found in [Williams90] and [Rasure91]. [Sauer90] describes how Khoros is used in the acquisition, processing and presentation of near ground level (NGL) data and its corresponding spatial characteristics.

7.4 Benchmarking and Validation

Benchmarking the performance of non-graphical hardware is an established field. Work has also been done on workstations. [Gallop90] reports on the evaluation of computational and graphical tasks on superworkstations and evaluation has also been reported on at a Eurographics Tutorial [Binot90]. When benchmarking graphical software it is common to require major rewrites of code however the task is becoming easier with the advent of common interfaces such as PHIGS PLUS.

If one considers benchmarking a visualization system, one has to ask what it is one wants to test, as problems are solved in different ways on different visualization systems.

Users may wish to know how quickly and effectively they can mount a specific problem on a given visualization system. Commonly a user's data is expressed in a form which does not correspond to any of the formats recognised by current visualization systems. They may also wish to know how quickly alternative display methods can be invoked. To solve this, some candidate problems would have to be posed.

A more mechanical approach is to test the how quickly the visualization system (the software and the hardware) can perform a particular display technique; for example volume visualization of a given data set using an isosurface technique. The data would be in a format most appropriate to the particular visualization system.

Validation in visualization systems is at present neglected. How does the user have confidence that the output from the visualization system is a correct representation of the data it has been given? Clinicians will not use volume rendered images which contain artefacts resulting from the algorithms. As far as the workshop participants were aware, no reported work has taken place on this problem.

7.5 Future Trends

The application builder systems described here are in their infancy, in functionality, available machine ranges and take up. It is clear that this situation will improve over the next few years.

There is at present little reported practical experience of the use of application builders in real applications and this needs to be remedied.

These software systems are very complex and validation of the basic algorithms used is a major problem. This will be particularly acute with those systems that are available cheaply, but unsupported.

The potential for volume rendering is increasing and there are a few software systems which provide this facility but mostly have no specific hardware assistance. Systems such as Pixel Planes from AT&T go part way to providing this feature and it is likely that increased use and understanding of parallel processing systems will improve this situation.

Massively parallel systems will be an important source of data for visualizing and also as a vehicle in which some of the visualization algorithms could execute. These systems are maturing in their compiler technology but still lack good graphics and visualization tools and systems.

It could happen that future improvements will come about by an increase in the number of processors available on superworkstations but not at the expense of software, so the increase from that direction is expected to be gradual.

Higher speed network access to the supercomputers joined with maturing "virtual reality" systems provides a very exciting next few years.

7.6 Conclusions

Three different categories of visualization software can be identified: subroutine libraries/presentation packages, turnkey visualization systems and application builders. These provide, in that order, increasing sophistication and functionality, but require increasingly expensive hardware to run effectively.

Subroutine libraries and packages provide very great scope and flexibility for the development of visualization systems, but incur possibly enormous cost to the user in writing and maintaining the programs. Many successful products have already been developed from them and they are widely used.

Turnkey visualization systems remove much of the load from the user by providing most of the program components but at a cost of reduced flexibility and extensibility. Most current commercial visualization packages use this type of approach, as reflected in the applications described in the application chapter. Many products are targetted at very specific applications.

The latest generation of software to be developed are the application builders. These are still relatively new and there are no mature products. They do however offer great potential as they appear to offer all the flexibility and extensibility of the subroutine libraries with the ease of use of the turnkey systems. Many new developments are expected in this form of product in the next few years.

At present though, there are few example systems which are widely available and which do not require high cost hardware to run. This does appear to be an area where any new work could be most effective. This work should deal with evaluating the systems available and also the hardware configurations, possibly distributed, which can most efficiently or economically support it. The implications on network access and bandwidth of these possibilities should not be forgotten.

It is also recognised that many parallel computer architectures could offer significant advantages when applied to visualization, data manipulation and graphical rendering. Some new work to research the problems of migrating visualization software to these systems should be considered, possibly involving collaboration with industry. Some small-scale work on this is in progress [Hubbold90].

7.7 Key References

Most references are provided in the bibliography at the end of the book. A few references which are considered key for this chapter are provided here.

[AGOCG90]

"PC Graphics Evaluation", *AGOCG Technical Report 3*, **(AGOCG, 1990)**.

A report of a survey on PC Graphics Presentation packages carried out by a graphics working party, on behalf of the UK Academic Community. This is available from Dr.A.Mumford, Computer Centre, Loughborough University, UK.

[Anderson89]

Anderson H.S., Berton J.A., Carswell P.G., Dyer D.S., Faust J.T., Kempf J.L., Marshall R.E., "The animation production environment: A basis for visualization and animation of scientific data", *Technical Report, Ohio Supercomputer Graphics Project*, **(Mar 1989)**.

A technical report from the Ohio SuperComputer Graphics Project which developed apE, the animation production Environment. It provides a general overview of apE and the technical background to its design. It is necessarily outdated in respect of the latest releases, for which see [Dyer90].

[Collins90]

Collins B.M., Phippen R.W., Quarendon P., Watson D., Whitfield G.A., Williams D.W., Wyatt M.J., "WinVis90 or a Mathematical Visualiser", *IBM UKSC Report No 227*, **(Apr 1990)**.

A report on WinVis90.

[Charalamides90]

Charalamides S., "New Wave technical graphics is welcome", *DEC USER*, **(Aug 1990)**, *pp 49-50*.

An article on PV-WAVE.

[Dyer90]

Dyer D.S., "A Dataflow Toolkit for Visualization", *IEEE Computer Graphics and Applications* vol 10 (4), **(July 1990)**, *pp 60-69*.

An article about apE 2.0.

[Jern89]

Jern M., "Visualisation of Scientific Data", *Computer Graphics 89,* ISBN 0-86363-190-3, **(Blenheim On-Line, 1989)**, *pp 79-103.*

An article on UNIRAS.

[Rasure91]

Rasure J., Argiro D., Sauer T., Williams C., "A Visual Language and Software Development Environment for Image Processing", *International Journal of Imaging Systems and Technology,* **(1991)**.

An article on Khoros, including the user interface development tools and the Cantata visual programming language.

[vandeWettering90]

van de Wettering M., "The Application Visualization System - AVS 2.0", *Pixel,* **(July / Aug 1990)**.

An article about AVS 2.

[Vandoni89]

Vandoni C.E., "Development of a Large Graphics-based Application Package", *Computers and Graphics* vol 13 (2), **(1989)**, *pp 243-252.*

An article on PAW.

Chapter 8

CONCLUSIONS

Edited by Rae Earnshaw

8.1 Summary - and where to get further information

This volume has presented a distillation of current techniques and applications of scientific visualization. The coverage is not exhaustive, but the tools and techniques that are described are representative of current practices in these areas. Further work is continuing on many of the topics highlighted in this volume.

A number of journals and societies are addressing the topic of scientific visualization, both from a theoretical and a practical point of view. Interested readers should consult the relevant literature such as IEEE Computer Graphics and its Applications, ACM Computer Graphics (and the Proceedings of the annual ACM SIGGRAPH conference), and the Proceedings of the recently established (1990) Visualization Conference organized by the IEEE Computer Society Technical Committee on Computer Graphics. The Visual Computer (published by Springer-Verlag) is the journal of the Computer Graphics Society (CGS) and publishes papers in the areas of computer vision, graphics, imaging, and applications. It includes detection and communication of visual data, intermediate data structures and processing techniques for visual data and computer graphics, and graphical representations of images. The Journal of Visualization and Computer Animation (published by Wiley) started in 1990 and publishes papers in the areas of animation and visualization techniques. Computer Graphics Forum is the journal of Eurographics and presents papers on the theory and practice of computer graphics. Its primary objectives are the dissemination of research results and of engineering developments to academic and industrial groups and individuals; reporting standardization activities; and reporting on events related to computer graphics. The British Computer Society has a special interest group on Computer Graphics and Displays that covers the general areas of vision, design, computer graphics, displays, human factors, and applications.

A further journal, the Supercomputing Review, contains useful information on aspects of supercomputing. The journal Computers and

Graphics (published by Pergamon Press) presents papers on computer graphics.

8.2 Methodology and Reference Model

In this volume, methodologies have been examined and a conceptual framework has been proposed. This should be regarded as early work in this area, and much remains to be done to relate this framework to other computer graphics reference models.

8.3 Techniques

A survey and classification of techniques has been presented to enable the reader to appreciate the technique being used and also its domain of applicability. A structure into which different techniques can be fitted has been proposed: this structure is a two-dimensional matrix - one axis indicates the type of entity being displayed (namely a set of points, or a scalar, vector or tensor field); the other axis indicates the dimension of the domain over which the entity is defined. Many of the common visualization techniques can be classified in this way and a broad sample of the common techniques have been described.

Each technique can typically be modelled as three separate processes: a model building step, in which a continuous model is constructed from the discrete data; a step in which the best means of visualizing the model is selected; and finally a rendering step. Traditionally the model building step has been the construction of an interpolant to fill out the data. Today however the increasing prevalence of large quantities of data to be analysed is focussing more attention on filtering and smoothing of data rather than on interpolation.

The current trend is to embed these techniques in an integrated, multi-functional system (the "application builders" described in Chapter 7: Products), and so some related issues have also been discussed: image processing, animation, interaction and perception.

8.4 Data Facilities

Handling large volumes of data is a theme common to both the Techniques chapter and also the Data Facilities chapter. This has increased the strategic importance of data standards, data exchange utilities, and data management. HDF and netCDF are examples of current work in this area and the wide interest in them is indicative of the real need for adequate and efficient data handling tools. Commercial database suppliers are increasingly targetting products at the scientific market, in order to provide solutions to current needs and expected future requirements.

A multiplicity of current data formats is in use, many of which have been summarised. Categories of data formats have been outlined, e.g. generic, application-specific, and image. However, it is unrealistic to expect all existing data to be converted for input into a visualization system. What is needed is the ability to handle the more common formats such as CGM, GIF, TIFF, HDF and PostScript. Toolkits for data format conversion are becoming increasingly available (e.g. PBMPLUS and SDSC) although further work is needed in this area.

Agreed standards for data transfer are becoming highly desirable, particularly in the area of data compression.

Because of the way in which standardization tends to arise from discipline-specific areas, the functionality and domain of the resulting standards can be overly restrictive. At the same time, data crosses all disciplines and wider coordination and collaboration across disparate areas will show clear advantages. Here are some examples of data standards that have arisen from particular constituent areas, but are nevertheless of relevance to a wider field: CGM (computer graphics metafile), raster interchange formats (image processing), STEP (CAD/CAM data exchange), ASN.1 (encoding standards). Visual computing tends to bring together areas that have been historically separated such as computer graphics and image processing, graphics data and modelling data. Hopefully the current work in scientific visualization will encourage greater inter-disciplinary efforts that will be to the scientists' mutual advantage.

Image storage and picture compression standards such as JPEG are becoming increasingly important as the use of video becomes the norm for interchange of data and results. Such standards are already being embedded in products (e.g. the colour NeXT workstation). Other workstation vendors are also moving in this direction.

8.5 Human-Computer Interface

Visualization comprises not only tools and products of the type outlined in this volume, but the end users' problem solving skills.

Designers of visualization tools should take greater account of research in cognitive science when devising frameworks and interface concepts for their systems. Representation and manipulation of multi-dimensional data is one specific area where this could be beneficial.

There is evidence that current visualization systems do not follow already known guidelines for designing user interfaces. Visualization may pose additional problems, such as choosing appropriate colour scales, or mixing colour and shading cues. Empirical studies are needed to formulate guidelines and reliable standards in these areas.

Visualization systems can change fundamentally the way that problems are solved and information is communicated. Organizations need to plan consciously for the introduction of such systems.

Flexibility is a crucial feature of visualization - the ability to try new ways to present and think about data. Thus visualization systems should not comprise a fixed set of tools, but should be open-ended and easily accessible to scientists.

Many current visualization products are based on window systems which were not designed with visualization systems in mind. These window systems may inhibit the development of more novel styles of interface, and incorporation of new types of device.

New styles of interaction are important and attention should be paid to these. For example, direct manipulation of user models is one area where visualization systems may be fundamentally different from traditional graphics or window system approaches. Studies are required to find what kinds of interaction are most useful for visualization, and to devise software systems which can support these.

8.6 Applications

At present, applications are using special purpose solutions and software. The more general purpose packages have yet to replace these. This is probably due to the current investment of time and effort in particular applications areas to develop specialized programs which do what is currently required. These become the base lines from which further developments are made. Moving to a more general software platform will probably not be perceived to be advantageous until the following conditions are satisfied -

1. Easy access to high power computation facilities so that there are no performance or interaction limitations.

2. Easy migration paths to enable users to specify their required functionality using general purpose software.

3. Successful use of visualization tools by pioneering scientists in particular disciplines, so that use of them spreads because of their reputation and functional advantages.

4. Easy to use facilities for exploiting the new hardware facilities that are increasingly becoming the norm, e.g. producing video output, utilizing multimedia interfaces, and effective utilization of super-computer facilities across networks.

Techniques and products concentrating on 3D visualization are readily available, since this is commonly required across a number of application areas, which are thus well catered for.

Other areas such as 2D data sets with variable visualizations and data sets with a larger number of dimensions are less well covered, and further developments are required in these areas.

8.7 Products

The majority of current visualization systems are what have been referred to in this book as turnkey applications. They provide an improved user interface and in many cases a good match with certain specific applications. However, the user interface is often not very flexible nor extensible.

Application builders offer much improved flexibility and the promise of being applicable to a wide range of applications. However, at present most practical uses of visualization centre around turnkey systems, or one-off packages that are built out of component parts.

Several developments can be anticipated:

1. Application builder products hold considerable promise, but their coverage of visualization techniques is sparse when compared with the sum of application needs. Exposure to real applications is needed in order to build up experience and an improved tools set.

2. Although most of the application builder products allow modules to be distributed, this capability is not well proven in real applications. Practical considerations of resource distribution and cost give rise to the need to carry out trials on specific practical distributed combinations.

3. Some components supporting visualization are becoming available on parallel hardware, but there are few signs at present that a complete visualization system can be easily tailored to exploit the capabilities of parallel hardware.

4. Validation of a visualization system is vital, but has hardly been considered in a systematic way. How do we know that that the visualization system is producing the correct response, or providing correct information?

5. The extent to which application builders are easy to use has yet to be evaluated.

8.8 Infrastructure Support

The current High Performance Computing and Communications Initiative in the USA has been formulated to provide a major upgrade in the computational resource available to scientists and engineers. The objective is to provide an enabling resource for previously unsolved problems in disciplines such as chemistry, meteorology, and astronomy.

One such proposed project is the EOS earth observation satellite system which will produce 10^{12} bits of information every day. A typical supercomputer centre in the USA currently has only built up around two terabytes of information in its user files in the past five years. Thus projects such as EOS will generate data at a rate considerably greater than current experience! In order to facilitate network access to supercomputer facilities to provide the resource to process such data, a National Research and Education Network is proposed. This would consist of fibre optic cables capable of carrying information at rates over 1 megabit/second.

Such considerations in the area of infrastructure support need to made in other countries on a national basis, in order to be able to plan for the facilities that are required.

8.9 Other uses of Visualization Tools

Tools such as those described in this volume can be also used by those whose primary interest is not in the scientific content of the information presented, but rather the creative or aesthetic value. Barlow et al [Barlow90] outlines how artists create effects and explores issues at the interface between art and science.

Artists and sculptors have been using computer-assisted tools for a number of years [Lansdown89] and these tools often promote new and unexpected ways of creating and developing images and objects. Thus visualization tools are not confined to scientific visualization but can be used in all areas where the user is seeking to create and manipulate information via visual means.

8.10 Virtual Reality Systems

Virtual reality is a set of hardware and software technologies to enable the user's body and senses to be translated into the information space to be examined. For example, a data glove can enable a user's hand to be projected into a 3D environment. By manipulating the glove the user interacts with the virtual world and can "handle" or "move" objects and issue commands. Alternative forms of interaction include body suits and data helmets with built-in stereoscopic viewscreens to enable the user to "enter" the virtual world. It is clear that these environments offer considerable potential in applications where the user has to interact in real time with objects in the viewing space, or develop an understanding of spatial relationships, or receive training in some particular procedure. Such techniques could be practised in the virtual world without the risk of expensive and dangerous mistakes. As such, they are developments and generalisations of procedures and techniques that have been used in flight simulators for years. However, flight simulators are expensive and usually only available to corporations such as airlines with the revenue to afford them! Current trends in visualization systems are to provide similar functionality but at much lower cost and for more generalised applications.

Such systems are capable of displaying large amounts of 3D information. It is possible to conceive of a scenario where a scientific visualization system interfaced to the user in a way that most suited the user's requirements. Such requirements could be to explore large data sets; look for similarities in structure or appearance of objects; investigate spatial relationships; or run simulations and observe the results directly.

Accessing information in large data sets has analogues with hypertext systems, where the user is provided with facilities for exploring data often using multimedia tools. It is anticipated that the development of interfaces for multimedia systems will also have benefits for visualization systems. Fairchild et al [Fairchild88] describe the use of a 3D navigator to access a large hypermedia network and display the results.

8.11 Importance of Scientific Visualization

The increasing occurrence of large volumes of data, high-powered workstations, new styles of interaction and user interfaces, fast networks, and new techniques for data display are all changing the nature of computing and what the user can expect from the tools now available. Scientific visualization encompasses all these areas and significant developments are expected in the future - [Thalmann91] and [Upson91].

8.12 References

[Barlow90]

"Images and Understanding: Thoughts about Images, Ideas and Understanding", ed. Barlow H., Blakemore C., Weston-Smith M., **(Cambridge University Press, 1990)**.

Outlines how artists create effects and explores issues at the interface between art and science.

[Fairchild88]

Fairchild K.M., Poltrock S.E., Furnas G.W., "Semnet: 3D graphics representations of large knowledge bases", *Cognitive Science and Its Applications for Human Computer Interaction*, ed. Guindon R., **(Lawrence Erlbaum Associates, 1988)**.

Aspects of cognitive science and its applications in human-computer interaction, with case studies of applications.

[Lansdown89]

"Computers in Art, Design and Animation", ed. Lansdown R.J., Earnshaw R.A., **(Springer-Verlag, 1989)**.

A collection of papers in the area of creative uses of computer graphics and associated tools and techniques.

[Thalmann91]

"New Trends in Animation and Visualization", ed. Thalmann D., Magnen-at-Thalmann N., **(John Wiley, 1991)**.

Computational and graphical techniques that are necessary to visualize objects and shapes are surveyed. It includes object oriented techniques, shape interrogation, virtual environments, animation techniques, and applications.

[Upson91]

Upson C., "Volumetric Visualization Techniques", *State of the Art in Computer Graphics - Visualization and Modelling*, ed. Rogers D.F., Earnshaw R.A., **(Springer-Verlag, 1991)**.

Survey and review of current techniques in visualization. The volume also contains a number of contributions in the areas of modelling, human-computer interface, parallel hardware, and image generation.

Appendix A

ENABLING TECHNOLOGIES

Complete visualization systems have been described in "Products" on page 175. Typically a visualization system may include a number of enabling technologies of which some examples, both hardware and software, are presented here.

A.1 Hardware

A.1.1 Hardware platforms for ViSC

Until recently the only way to process complex or voluminous data sets was using supercomputers, such as the Cray, then passing the processed data over to a graphics terminal or workstation for display. As more specialized graphics hardware became available, an increasing amount of the graphics processing could be left to the workstation, hence reducing the volume of data transferred over the network connection. Despite this reduction, the data to be processed appears to be increasing so rapidly that even with improved network speeds it is not possible to keep up with demands. Supercomputers are very effective at the "number crunching" required in initial processing or simulation of data. However the cost involved in purchasing and maintenance leaves them out of range for most budgets. Even for departments where such a system is affordable, the system would generally be utilised to its full by sharing access time between several projects and possibly imposing CPU usage and disk quotas - this may reduce the likelihood of repeated processing with slightly modified data.

More practical, financially, for most departments would be a system which could be expanded as and when money and expertise becomes available. Transputers are one such solution where hundreds or thousands of processors can be linked together to provide the necessary computing resources for visualization. The complexity of programming such systems have caused major delays in the appearance of software products, including stable language compilers. Generation of ray-traced images was one of the first graphics applications to take advantage of these highly parallel architectures, but without stable compilers general purpose graphics software has been slow to materialise. To date most, if not all, transputer-based graphics and visualization software is very much application-dependent.

Another approach is to use hardware that goes part way to providing the power of a supercomputer but has tightly coupled high performance graphics hardware, e.g. Alliant VFX models (the Visualization series), Hewlett-Packard's "Apollo" DN10000, Silicon Graphics PowerSeries and PowerVision machines, Stardent ST2000 (formerly Stellar GS2000) and ST3000 (formerly Titan 3). A descriptive name commonly used for this type of hardware is "superworkstation". Typically, superworkstations provide 10% of the computational power of a Cray supercomputer at only 1% of the cost, or at the other extreme ten times the power, both in computational and graphics performance, of an average Sun workstation at approximately four times the cost.

To achieve the performance required to process, display and interact with complex and large volumes of data, superworkstations provide hardware for vectorization, or a small number of parallel processors (typically four or eight), or both facilities. Despite the small number of parallel processors involved, parallel compilers are not currently available for all the above mentioned hardware platforms, and even where such compilers exist, problems such as bugs and poor utilisation of the processors are still apparent. These problems have not prevented the majority of visualization products being targeted at the superworkstation market.

Systems such as the forthcoming Colour NEXT workstation show promise in the area of video and multimedia and have some of the capabilities of the 3D workstations also.

A.1.2 Graphics input for ViSC

The type of input device commonly required for use with visualization systems are those that enable the user to interact with a graphical representation of their data. In the future it will be necessary for visualization products to be able to process data directly from data sources (such as telescopes, x-ray devices, digitisers, etc.) and hence these will also be classified as input devices but for the present interaction devices will be the main focus.

Mouse (optical/mechanical)

The mouse has been successfully used for interaction with 2D applications for some time now. It is also possible to use it with interactive 3D graphics. There are two approaches to using the mouse as a 3D input device [Bier90]:

1. divide the 3D operations into a sequence of 2D operations (for example, select the current axes pair from xy, yz, and zx and perform all the 2D transformations relative to that).

2. map 3D operations onto the mouse buttons. On a typical 3-button mouse where one button is reserved for selection or menu operations, the second could indicate translation while the third indicates rotation. Two-dimensional interaction can be mapped with one mouse button by dragging the mouse forward and backward, or left and right while the third dimension can be mapped by the use of a simultaneous keyboard character (typically "shift", "alt", or "meta" keys).

Dial/Knob boxes

Typically these consist of eight dials which can be used to modify eight parameters to an application. In addition to being able to translate and rotate in X, Y and Z it would be possible to interactively change two further properties, such as light intensity and transparency or for more application oriented input. The problem with this device is that it is difficult to modify more than one input at a time so it is not possible, for instance, to rotate around x,y and z simultaneously. [Beaton87]

Data Glove

Data gloves [Blanchard90] use various sensor technologies to interpret hand position and orientation in 3D-space along with hand gestures. For example, finger flexion may be detected by mechanical means, by using materials of various electrical resistance (greater resistance when bent), and lossy fibre optic cables (light loss proportional to bend). The information that follows refers to the commonly used VPL product called the "DataGlove".

Simple techniques have been developed to interact with objects, such as pointing a finger in the direction of an object to select it, closing the hand to pick up the object, tracking hand position to map new positions and orientations of selected objects (and/or the 3D representation of the users hand), then opening the hand to indicate the object is to be released. These techniques are particularly useful in visualizations relying on the user becoming a part of a 3D "virtual world". Some problems observed include:

- current software and hardware are not always able to keep up with the user's gestures;
- a calibration session is required to tailor the glove to each individual user;
- during long interaction sessions it is possible that the user's arm may weary;
- the sensors attached to the glove could come adrift during movements of the hand;
- since the DataGlove is connected by cables, to the hardware processing the data, the users movement is restricted;
- since the electromagnetic field is set up by a fixed emitter the glove operator needs to be within a certain range of the data receiver (around a few feet, hence restricting the length of the connecting cables).

Spaceball

The Spaceball, from Spatial Systems Inc., resembles a trackerball in appearance except the ball does not actually move, but simply senses the forces applied [Spatial89a], [Spatial89b]. The spaceball has the added advantage of being able to sense torque thus enabling rotation, at the same time translation, to be represented. Since the user nor the device has to move, the movement of objects are not restricted by hardware as in the case of the DataGlove. To continue animation of objects simply requires the user to persist with the desired pressure. It appears to be an easy to use device, with a little practice. Sensitivity can be modified to adjust to the individuals needs. As the device is small and remains stationary during operation, unlike a data glove, it is ideal for demonstration purposes where interest should be on the display and not on the operator. This device appears to be more suited to screen interactions (typically 2D representations of 3D data) rather than for use in 3D "virtual worlds".

3D locator

A number of input devices are beginning to appear which allow more than three degrees of freedom. At the Eurographics UK conference in 1989 an input device was demonstrated with six degrees of freedom. It was essentially a three-dimensional locator which could also provide pitch, yaw and roll. The demonstration (on an Archimedes) was a three-dimensional gyroscope which could be located, scaled and oriented using a single stylus-like pointer. Work on a six-dimensional, one button has also been reported - this is called a bat [Ware88].

A.1.3 Graphics output for ViSC

Typically output from 3D applications are as 2D images, e.g. on the screen or as hardcopy in the form of photographs, slides, viewfoils and on paper. This section discusses the generation of 2D images on colour hardcopy and also a number of other novel techniques which are becoming available.

Colour Hardcopy

The requirement for obtaining hardcopy of the pictures generated on the screen quickly, conveniently and in colour should not be forgotten when costing, evaluating or purchasing visualization software. Viewfoils and paper can be generated by a variety of colour thermal, inkjet and plotting devices at a wide range of costs and image quality. Hardcopy can best be obtained via data export methods such as Postscript or CGM. Where a picture resulting from a visualization session goes beyond the geometric capabilities of CGM or Postscript, then recourse may be had to outputting an image, which is possible but not ideal with these formats. Photographs, directly from a screen, are also used.

Large format colour pictures can only be supplied by electrostatic or ink-jet printers and are expensive to purchase and to maintain and run. Small format colour devices (A3 or A4) are much cheaper to buy and can use ink-jet, thermal wax transfer or thermal dye diffusion technologies. Running costs are proportional to quality, with ink jet systems the cheapest and dye diffusion the most expensive (a factor of ten can separate the two).

Head-Mounted Displays

A pair of stereoscopic images are fed to the user to give the impression of being a part of the image, e.g. virtual worlds [Wright89]. The user views the virtual reality through eyeglasses in the form of 3D stereo pictures, movement of the head in one direction causes the image to be moved in the opposite direction to compensate. Used in conjunction with a data glove or some other 3D interaction device, this may be a part of the visualization system of the future. In [Fuchs89b] the suggestion is made that this display with some hand-held positioning device could be used to investigate volume data, including being able to move inside the data. The hand-held device could then be used to interactively delete areas of no interest or to uncover hidden features. One system, under development, uses the head mounted display with an eye-tracking device to provide gaze directed volume rendering (see [Levoy90a]).

Video Products

The easy availability of video recorders and their ability to present full colour, full-speed animation output to any number of people has made video a popular form of visualization output. Systems for the production of video output have become more common as broadcast television and personal computing have simultaneously latched onto computer graphics techniques.

Video may be recorded in a number of ways.

- Output produced on a computer screen may be transcribed to videotape at the rate it is produced; if the computer screen is not of the domestic encoding format (PAL or NTSC compatible) an intermediate standards convertor or equivalent is required.

- Output may be rendered a frame at a time into a frame-store and, by use of an animation controller and an editing recorder, may be stored a frame at a time onto videotape.

- Output may be produced a frame at a time and recorded onto a magnetic (rewritable) or optical (normally Write Once) disk and subsequently copied from disk to tape if required on tape.

- Output may be stored in RAM or fast disk and replayed rapidly to a screen; in extreme cases (such as the UltraNet system) the replay is actually over a very high speed network).

The performance of the recording system and the quality of the output depend crucially on the cost of the system. Full digital broadcast quality and the flexibility of variable speed transcription from disk to tape is achievable with disk systems such as those from Abekas. For a fraction of the cost, frame-at-a-time editing to tape can produce semi-broadcast quality but will require around 30 seconds recording time per frame.

Worldwide standards for video appear no closer now than at any time in the past. Although a long planning period for High Definition TV provided the opportunity for solving the PAL / NTSC / SECAM and 50 Hz / 60 Hz incompatibility problems, the current proposals from Japan, the USA and Europe are all different and as difficult to convert as at present.

High Definition TV does offer a hope that the premium price currently paid for high resolution colour graphics displays may be reduced if and when High Definition TV becomes a domestic reality. Equally, it will provide a new generation of video recorders capable of storing and replaying pictures at the quality of current high resolution displays.

It is interesting to note that the current interest in (and rapid development of) image compression systems has been forced by the research into High Definition TV systems, since the bandwidth for transmission of uncompressed high resolution pictures was not available.

Other Techniques and Media

Animation, depth-cueing, lighting, shading and shadows, and more recently the use of stereo monitors with special glasses, such as those from CrystalEyes [Robinson90], can improve the appearance. Holograms can be used to produce high quality static 3D images. Genuine 3D devices have been reported from time to time, including a 3D multiplanar spinning disc [Williams88] and devices which effectively cut and shape a 3D object.

A.2 Graphics software for ViSC

This section describes graphics software that can be used to support visualization systems. They do not provide visualization capabilities as such.

A.2.1 2D Graphics Systems

Where the visualization system is creating a 2D abstract visualization object, it can be appropriate to use a 2D graphics system for interaction and display. The standard in this area is GKS [ISO(7942)85], with language bindings for Fortran, Pascal, Ada and C [ISO(8651-1)88], [ISO(8651-2)88], [ISO(8651-3)88] and [ISO(8651-1)90].

A.2.2 3D Graphics Systems

Often, the visualization system creates a 3D abstract visualization object. Such objects can be drawn with a 2D graphics system with software to perform conversion between 3D and 2D. However a 3D graphics system allows effective manipulation of the 3D object on a suitable workstation.

GKS-3D is an ISO Computer Graphics standard for the display and manipulation of 3D primitives. Although similar in many respects to PHIGS, which is described below, it differs from PHIGS and shares with GKS the characteristics of a simple one level graphics storage scheme and allowing graphics to be output without first using graphics storage. The language bindings for Fortran, Ada and C are defined in [ISO(8806-1)88], [ISO(8806-3)88] and [ISO(8806-4)90].

PHIGS is an ISO Computer Graphics standard [ISO/IEC(9592-1,2,3)89], for the modelling, display and manipulation of 3D primitives. It is a subset of PHIGS PLUS, which is described below, but does not have PHIGS PLUS's ability to define surface characteristics, lighting effects and output primitives consisting of multiple polygons or curved surfaces. Implementations are becoming available on a number of workstation vendors' hardware. Additional facilities which exist in PHIGS PLUS but not in the PHIGS definition are commonly, but selectively, provided. The language bindings for Fortran, Ada and C are defined in [ISO/IEC(9593-1)90], [ISO/IEC(9593-3)90] and [ISO/IEC(9593-4)90].

PHIGS PLUS / PHIGS+ Originally defined as PHIGS+ [ANSI88], this is being processed towards an ISO standard as PHIGS PLUS - [ISO/IEC(9592-4)91] and [ISO/IEC(9592Am1)91]. At present, most current implementations are written with the original PHIGS+ definition in view but this is expected to change. It is suitable for hierarchical models and those situations where the data sets change infrequently, but need to be rotated, translated, or have overall properties change (such as transparency or shading method), such as in CAD, but where rapidly changing objects are needed (such as deforming solids or fluid flow), the complete set of data needs to be stored and traversed for each snapshot. Recent changes to PHIGS PLUS have led to data being associated with objects; the user has control over the mapping of data to colour - data mapping. This recent change solves some problems with colour interpolation and, for some applications, the need to store the entire data structure for each snapshot is reduced.

PHIGS PLUS allows complex objects to be specified with only a single procedure call (creating a single structure element), but editing part of the structure element, which might be useful for debugging, is not possible.

PHIGS PLUS, like PHIGS, was not specifically designed for a window environment. Typically it depends on the implementation as to what control, if any, the user has over the creation, manipulation and deletion of windows and the possibility of interaction with that window (such as keyboard or mouse input). However, the use of X11 calls for windowing tasks is becoming popular in implementations.

Doré For visualization, this has some of the strengths of PHIGS PLUS - although Doré does not have the PHIGS PLUS data mapping capability - but allows immediate mode graphics too [Ardent88a]. For reasons of compatibility, PHIGS PLUS needs to allow PHIGS applications to run and therefore some of its defaults are not easy to use when lighting and shading are required.

Doré has more reasonable defaults for many features, which makes it easier for the first-time user to display objects with relatively little code or knowledge of Doré. By allowing variable data it is possible to represent fluid-flow and deforming objects.

Doré is becoming available on a wide range of products and is commercially available from Intelligent Light for platforms such as Hewlett Packard's Apollos and Silicon Graphics.

RenderMan This produces high-quality and possibly realistic images but not in real-time - [Pixar88] and [Upstill90]. It does have CSG primitives and operations which could be useful in CAD applications but does not have hierarchical objects. Using RenderMan for displaying objects is currently too time consuming to be considered for interactive visualization products.

GL This graphics library from Silicon Graphics provides convenient access to the accelerated graphics facilities on Silicon Graphics hardware, which includes facilities that go beyond PHIGS PLUS, such as texture maps. There are undoubted performance advantages when used on the hardware to which it is tailored. The disadvantage of lack of interworking associated with proprietary software (although GL is becoming available on other hardware) needs to be considered.

A.2.3 Window Systems

X11 Commonly available at version 11 release 3 and release 4, X11 is a windowing system freely available from MIT [O'Reilly88], [Nye88]. X11 provides 2D graphics facilities but also gives great control over the user interface of an application. Toolkits are becoming more readily available to assist in the standardization of these interfaces (for example, Motif). The Xlib library provides 2D graphics capabilities and is good for pixel level processing (suitable for image processing and volume rendering applications). As X11 is available at little or no cost, many academic establishments have been attracted to this software leading to availability on a wide range of hardware. In addition as its popularity increases more implementations are moving to hardware/firmware.

PEX [Clifford88] is an extension of X11 to allow workstations to support many PHIGS+ capabilities in firmware. Sun Microsystems are understood to be developing a sample implementation of PEX, which is expected to be made available at X11 release R5. There is also a freely available version UIPEX from the University of Illinois. Several companies are expected to provide the PEX specification in microcode and at least one company, Evans and Sutherland implemented an early version.

A user would not invoke PEX directly as a graphics library. Instead Application Programming Interfaces (APIs) such as PHIGS and PHIGS PLUS would be called and are expected to be supplied with PEX. The advantage to a user is that if their application program or a visualization system uses the PHIGS or PHIGS PLUS programming interface, improved performance should be possible on a workstation that implements PHIGS or PHIGS PLUS via a firmware implementation of PEX.

The current PEX document is obtainable online from expo.lcs.mit.edu in the U.S.A.

A.3 User Interface Toolkits

Most User Interface Toolkits are based on X11. The advantage is that such toolkits have the potential to become widely available, some particularly so. The inherent restriction is that the graphical input devices that have promise for visualization, such as the spaceball, would be excluded by a toolkit that conformed to the standard interface. Examples, with their associated user interface style, are as follows.

Toolkit	User Interface Style
OI	Motif and OPEN LOOK styles
Interviews	Interviews style
XView	Open Look style
Open Look GUI (also known as Open Look Intrinsics Toolkit - OLIT)	Open Look style
Motif	Motif style
XVT (eXtensible Virtual Toolkit)	several styles, including Motif, Open Look and Mac

Table 14. User Interface Styles for various User Interface Toolkits

A.4 Database Systems

A.4.1 Basics

The principles behind a database system are very simple and can be summarised by the following definition:

> *A database is a collection of well organized records within a commonly available mass storage medium. It serves one or more applications in an optimal fashion by allowing a common and controlled approach to adding, modifying and retrieving sets of data. The Database Management System (DBMS) is a suite of computer programs which perform these operations in a standardized and fully controllable manner. [Yannakoudakis88]*

It can be seen that a visualization system, in common with many other classes of computing system, can benefit from the organizational aspects of a database management system.

Date [Date81] provides a general introduction to database concepts. Many different database models exist but broadly speaking, they can be categorised (based on their architecture) as being of one of a number of different types:

- Hierarchical (or tree)
- Networked
- Relational
- Object-orientated

In the context of visualization systems, the relative advantages and disadvantages of the schemes are as follows.

Hierarchical

The advantage of these kinds of database is that they tend to be associated with fast performance and do not waste disk storage. In practice it is often difficult to model data using hierarchies and they can be very hard to restructure or tune for changing requirements.

Networked

The multiplicity of links can make it difficult to keep track of where, in the structure, we currently are and maintenance of the database can be complicated.

Relational

Implementations have had a tendency to be inefficient but with the advent of widely accepted commercial systems such as ORACLE and INGRES relational databases have become very widely used; this has necessarily forced improved efficiency.

Relational databases are not ideally suited to the storage of textual information (e.g paragraphs), pictures and attribute data (beyond simple lines of text).

Object-orientated

Object-orientated databases (OODBs) are quite similar to hierarchical or network databases. OODBs use a nested structure of objects, whereas hierarchical databases use a manually navigated record structure; both models use the notion of pointers.

In OODBs each object has both a state - the set of values of its attributes - and a behaviour - the set of methods which operate on the object's state. An object's state and behaviour are encapsulated so that it is visible only through its response to messages. This makes OODBs much more suitable for use with large, complex data structures than any of the other types of database model. However, the theory of OODBs has yet to be developed fully [Maguire90].

Broadly speaking there are three classes of database user:

Application Programmer

who wishes to utilise/manipulate a database from an applications program. Operations which they might wish to perform include the insertion of new information or the retrieval, updating or deletion of existing information.

End User

who wishes to manipulate a database directly, probably from a terminal, either by invoking an application program, a set of macros or using the facilities of a Data Manipulation Language.

Database Administrator

who maintains overall responsibility for the data contained in a database and performs general housekeeping facilities such as archiving.

A.4.2 Database Management System Products

The majority of commercially available database management systems are based on the relational model. Some of the most commonly used are INGRES, ORACLE and TekBase.

INGRES

INGRES is an integrated relational database system designed for commercial, scientific and engineering applications. Since release 4.0, INGRES supports two query languages, SQL and QUEL. SQL is popular for commercial data processing and QUEL is widely used in the scientific and engineering world. They are largely interchangeable, being based on the same underlying INGRES software routines. These query languages can be invoked from programming languages such as Ada, Basic, C, COBOL, Fortran, Pascal and PL/1.

Transparent access to INGRES databases on remote computers can be achieved by INGRES/NET.

INGRES has been implemented on a wide range of hardware, from PCs (where it has a LOTUS 1-2-3 style interface) to minis and mainframes.

Recently INGRES have announced their "Intelligent Database", a new version of their relational database, which provides control over knowledge management, object management and data management by introducing AI techniques to RDBMS technology. This effectively removes many of the traditional limitations inherent in RDBMS systems (e.g. inflexible data structures and limited data types).

ORACLE

ORACLE is a Relational Database Management System with an integrated set of software productivity tools. The capabilities of the database kernel are dictated by the facilities available in the ORACLE variant of SQL. Optional components such as SQL*PLUS and EASY*SQL are available which allow end users to access data in an ORACLE database, or PRO*SQL to develop applications.

ORACLE's implementation of SQL includes the following features in addition to those in the SQL standard.

- SET operators
- ability to add new data definitions and to alter existing data definitions
- dynamic SQL constructs
- support for additional data types

ORACLE currently support a number of products (SQL*Net, SQL*STAR and SQL*Connect) which provide distributed capabilities. It also supports interface products which allow programmers to develop their applications in high level languages such as C, Cobol, Fortran, Pascal, PL/1 and Ada.

ORACLE is available on PCs, minicomputers and mainframes.

TekBase

TekBase is described by its developers, Leading Technology Products, as a relational database management system for technical applications. It supports a range of scientific data types, including relatively large numeric arrays (128,000 points) and also supports the import of data directly from instrumentation and remote computers.

TekBase has its own query language TQL which is similar to but not compatible with SQL. TQL contains a large set of computational commands and mathematical functions; application-specific function libraries are available as options. Specialized facilities are included to allow the generation of graphs, performance of statistical analyses and document generation (with integrated graphics/text).

TekBase uses a modern client-server architecture to allow data to be stored in high performance data servers, located on a Unix network, with user access via networked personal workstations or terminals. It is currently available for the full range of Hewlett-Packard and Sun Microsystems computers for application development. TekBase can be used from C, Fortran, Pascal and Basic programs.

A.4.2.1 *Commercial DBMS Systems*

The following table documents the characteristics of a number of commercial DBMS systems.

Product	**ORACLE**	**INGRES**	**TekBase**
Database Query Language	SQL+	SQL, QUEL	TQL
Language Support	ADA, C, COBOL, FORTRAN, PASCAL, PL/I	ADA, BASIC, FORTRAN, C, COBOL, PASCAL, PL/I	BASIC, C, FORTRAN, PASCAL
Distributed Database Facilities	SQL*NET, SQL*STAR, SQL*Connect	INGRES/NET, INGRES/STAR	yes
Hardware Interfaces	PCs, minis, mainframes; >50% are DEC or IBM	PCs, minis, mainframes	Sun, Hewlett-Packard
Applications	Commercial	Commercial, scientific, engineering	Scientific
Supplier	Oracle Corporation	Relational Technology Inc.	Leading Technology Products

Table 15. Relational Database Systems

A.5 ViSC Generic Data Formats

HDF [NCSA89] and NetCDF ([Rew90a] and [Rew90b]) were introduced in chapter 4, as generic data formats. This section provides more specific information.

It has been announced that there are plans to investigate the inclusion of NetCDF in HDF, subject to financial backing for the project.

A.5.1 HDF

A.5.1.1 Architecture

HDF provides a general purpose file structure that encompasses the following:

- makes it possible to obtain information about the data (for example, a look-up table) directly from the file rather than another source;
- enables the storage of arbitrary mixtures of data and related information in different files, even when the files are processed by the same application program;
- standardizes the formats and descriptions of many types of commonly used datasets, such as raster images and scientific data;
- encourages the use of a common data format by all machines and programs that produce files containing a specific dataset;
- can be adapted to accommodate virtually any kind of data by defining new tags or a new combination of tags.

A.5.1.2 Grouped Sets

Related items of information about a particular type of data are grouped into sets (raster image sets, scientific datasets, etc.). Each set defines an application area supported by HDF. Additional sets can be defined and added to HDF as the needs arise.

- An 8-bit raster image set (RIS8) is a set of tags and associated information required to store an 8-bit raster image in an HDF file. It contains an image (a two-dimensional array of 8-bit numbers, one for each pixel in the raster image), dimensions (x and y), a compression scheme (run length encoding or IMCOMP) and optionally a palette lookup table with 256 entries, associating a colour with each of the 256 possible pixel values.
- A 24-bit raster image set (RIS24) contains at least the following components: an image (a two-dimensional array of 24-bit pixel representations, where each has three 8-bit components, one for each of red, green and blue; an interlace scheme that describes the order in which the pixel components are physically stored in the

file (pixel schema, scan-line schema or scan-plane scheme); dimensions (x and y).

- A Scientific DataSet (SDS) is an HDF set that stores rectangular gridded arrays of floating point data, together with information about the data. An SDS might contain: floating point data values (IEEE by default); the number of dimensions and their sizes; the coordinate system; the scales, labels and units for all dimensions and data; format specifications to be used for display; the maximum and minimum data values.

- Annotations are textual information concerning the collection, meaning or use of a file or data. This could include: titles, comments, variable names, formulae, parameters or source code.

- Vsets, a new feature in HDF, are designed for use by applications which use meshes, polygonal or connective data.

A.5.1.3 Access

HDF provides interfaces callable from Fortran and C. It also provides a set of command line utilities, which perform common operations without the need to write a program.

A.5.1.4 Availability

HDF currently supports sharing data across machines and systems such as CRAY (UNICOS), Silicon Graphics (UNIX), Alliant (CONCENTRIX), Sun (UNIX), VAX (UNIX), Macintosh (MacOS), and IBM PC (MSDOS).

A.5.2 NetCDF

A.5.2.1 Architecture

NetCDF software uses the concept of an abstract data type, via a defined set of C library functions. Internal data representations can be changed without affecting the user program. To achieve network transparency, netCDF is implemented on XDR [Sun88].

A.5.2.2 Data

The components of a netCDF file are:

- netCDF dimension - a named integer used to specify the shape of one or more of the multidimensional variables, contained in a netCDF file.
- netCDF variable - this represents a multidimensional array of values of the same type. A variable has a name, a data type and a shape described by a list of dimensions. It may also have data values and associated attributes.
- netCDF attribute - this is information about a variable or an entire netCDF file. This is analogous to the information stored in data dictionaries and schemata in conventional database systems.

The netCDF library allows access to the structure of the data held in a netCDF file.

The current set of primitive data types supported by the netCDF interface are: byte, character, short, long, float and double.

The netCDF interface supports direct access to single data values or to an arbitrary hyperslab of data for a single variable. A hyperslab is a multidimensional subset of the entire variable, and is specified by the indices of a corner point and a list of edge lengths along each of the dimensions of the variable.

The only kind of data structure directly supported by the netCDF abstraction is a collection of multidimensional variables with attached vector attributes. NetCDF is not particularly well suited for storing linked lists, trees, sparse matrices or other kinds of data structures requiring pointers.

NetCDF can be used as an archive format for storing data, but it will generally take more space than a special-purpose archive format that exploits knowledge of particular characteristics of a set of data.

A.5.2.3 Access

In addition to the library procedures, utilities are also available to generate a netCDF file from a defined readable form called CDL.

A.5.2.4 Availability

The netCDF software provides common C and FORTRAN interfaces for applications and data. The C interface library is available for many common computing platforms, including UNIX, VMS, MSDOS, and MacOS environments. The FORTRAN interface is available on a smaller set of environments (due to the lack of a standard for calling C from FORTRAN).

A.6 Key References

Most references are provided in the bibliography at the end of the book. A few references which are considered key for this chapter are provided here.

[Date81]

Date C.J., "An Introduction to Database Systems", **(Addison-Wesley, 1981)**.

A good introduction to the theory of databases, in particular highlighting the differences between the common database architectures; these are illustrated by reference to IBM products.

[ISO/IEC(9592-4)91]

ISO/IEC DIS, "Information processing systems - Computer Graphics - Programmer's Hierarchical Interactive Graphics System - Part 4 : Plus Lumière und Surfaces, PHIGS PLUS", **(1991)**.

The ISO definition (at present a draft) of PHIGS PLUS. This should be read with PHIGS [ISO/IEC(9592-1,2,3)89] and its amendments [ISO/IEC(9592Am1)91].

[Spatial89a]

Spatial Systems Inc., "Spaceball Technical Reference V2.0", **(1989)**.

This is the technical reference manual to one of the graphical input devices covered in this appendix, the spaceball.

[Yannakoudakis88]

Yannakoudakis E.J., "The Architectural Logic of Database Systems", **(Springer-Verlag, 1988)**.

Reviews the basic architectural types of traditional database (relational, hierarchic, network) and introduces associated terminology.

Appendix B

GLOSSARY

This appendix consists of two sections:

- **Visualization Terms** - words or phrases used with specific meanings in the context of visualization; included in this section are a number of computer graphics terms that are used throughout the book;
- **Abbreviations and Acronyms** - abbreviations and acronyms of products, techniques and organizations that are used throughout this book.

B.1 Visualization Terms

bi-cubic spline: a cubic spline (*q.v.*) applied in two dimensions.

connectivity: a relation between two voxels, which are said to be 6-connected (18-connected, 26-connected) if and only if there exists a six path (18-path, 26-path - see path) between them [Kaufman90].

contour: the (closed) boundary of a structure or the outline of a region in a 2D data plane or cross-section.

cross-section: the set of voxels residing on a plane that passes through the dataset.

cuberille: a representation of the volumetric dataset as a thresholded binary array of opaque cubes having six mutually orthogonal square faces [Chen85].

cubic spline: a form of spline (a piecewise polynomial function) commonly used to approximate natural curves [Ahlberg67].

cuboid: a six-sided each face of which is a rectangle.

database: a collection of well organized records within a commonly available mass storage medium serving one or more organization; each contains both a stored values and necessary control data to manipulate them.

database administrator: the person responsible for the overall organization and maintenance of the database and database management system.

database management system (DBMS): an integrated layer of software which enables the creation, manipulation, maintenance, or reorganisation of a database at all levels in a consistent, controllable and efficient manner.

data description language: a programming sublanguage, part of the DBMS package, used to describe and define data and relationship at the logical level prior to any processing which may take place.

data manipulation language: a programming sublanguage, part of DBMS, which enables application programs to access and manipulate data.

data type: the concept of data type is closely related to the concept "domain" and ultimately defines the set of values which a variable can take. A variable may represent a data item or an attribute within an application program or database environment.

depth-cueing: the variation of the intrinsic colour of an object according to its distance from the virtual viewer.

dividing cubes: an algorithm that subdivides voxels that lie on the surface of the object into a small cubes which are represented as point primitives (cf. marching cubes) [Cline88].

filter: an algorithm which is applied to a set of data to select a subset.

glyph: a character, marker, symbol or icon.

interpolation: the process of computing new intermediate data values between existing data values.

isovalue surface: a pseudosurface of constant density within a volumetric dataset [Purvis86], [Shirley89].

marching cubes: an algorithm that extracts an isovalue surface from a volumetric dataset by a sequential tessellation of the boundary voxels by tiny triangles (cf. dividing cubes) [Lorensen87].

metadata: data which describes and defines other data.

multi-modal visualization: the simultaneous visualization of multiple datasets of the same object or phenomenon obtained through the use of several sampling technologies [Hu89].

neighbourhood: the 6-neighbourhood (18-neighbourhood, 26-neighbourhood) of a voxel is the set of all voxels that are 6-adjacent (18-adjacent, 26-adjacent - see adjacency) to it [Kaufman90].

nominative data: designates (or names) different types of entity [Foley90b].

NURBS: non-uniform rational B-splines, a powerful way of describing arbitrarily complex 3-dimensional surfaces, used in PHIGS PLUS [ISO/IEC(9592-4)91].

Nyquist frequency: the lower bound to the frequency at which a signal can be sampled such that its original form can be reconstructed from the samples [Foley90b].

octree: a tree data structure that represents a volumetric dataset by recursively subdividing it into octants. Octree is the 3D counterpart of the quadtree [Meagher84], [Samet88a].

opacity: the relative capacity of a material to obstruct the transmission of light through it (cf. translucency, transparency).

ordinal data: is ordered but no metric information is associated with it [Foley90b].

path: a sequence of voxels is a 6-path (18-path, 26-path) if every two consecutive voxels along the sequence are 6-adjacent (18-adjacent, 26-adjacent - see adjacency) [Kaufman90].

particle tracing: a technique of visualizing flows by drawing the trajectory of particles released within the flow.

pixel: an abbreviation for "picture element".

quadtree: a way of subdividing areas, used as a way of storing images [Warnock69].

query language: a sublanguage, part of DBMS, which enables the interrogation of databases.

ratio data: is both ordered and has an associated metric [Foley90b].

ray casting: in volume rendering, a volume viewing algorithm in which sight rays are cast from the viewing plane through the volume. The tracing of the ray stops when the visible voxel are determined by accumulating or encountering an opaque value (see forward projection) [Levoy88].

ray tracing: in volume rendering, a volume viewing algorithm in which light behaviour is simulated by recursively tracing individual imaginary rays of light through the scene [Webber90].

shading: a colouring of the image pixels that represents the light intensity transmitted to each pixel, taking into consideration the light model, e.g. direct and indirect illumination, position of the light sources and emitters, and the position and orientation of the image [Tam88], [Tiede90].

surface rendering: an indirect technique used for visualizing volume primitives by first converting them into an intermediate surface representation (see surface reconstruction) and then using conventional computer graphics techniques to render them (cf. volume rendering) [Kaufman90].

tiling: a method that converts a stack of contours into a surface that is represented as a polygon mesh by connecting adjacent contours with polygonal tiles [Fuchs77].

transfer function: a function that maps the sampled voxel values into presentable parameters such as colour, intensity, or translucency [Upson88].

transparency: the property of a material which allows the transmission of light through it without appreciable scattering so that objects beyond are entirely visible (cf. opacity, translucency).

thresholding: a technique used primarily with surface rendering, in which a density value of the interface between two materials in the dataset is selected so that the interface surface can be identified for rendering [Herman83].

triangulation: a tiling method in which the tiles are triangles [Keppel75].

tri-cubic spline: a cubic spline (*q.v.*) applied in three dimensions.

ViSC: abbreviation for "Visualization in Scientific Computing".

visualization: a method of extracting meaningful information from complex datasets through the use of interactive graphics and imaging [McCormick87].

volume rendering: a direct technique for visualizing volume primitives without any intermediate conversion of the volumetric dataset to surface representation (cf. surface rendering) [Drebin88], [Levoy88].

volume shading: the technique of shading a volumetric dataset or its projection.

volume visualization: a visualization method concerned with the representation, manipulation, and rendering of volumetric data.

voxel: an abbreviation for "volume element" or "volume cell". It is the 3D conceptual counterpart of the 2D pixel. Each voxel is a quantum unit of volume and has a numeric value (or values) associated with it that represents some measurable properties or independent variables of the real objects or phenomena.

B.2 Abbreviations and Acronyms

ACM: Association of Computing Machinery.

ADCT: adaptive discrete cosine transform; a technique for data compression (*see "Compression Techniques" on page 101*).

AGOCG: Advisory Group On Computer Graphics; joint group between UK Research Councils and UK Computer Board responsible for the coordination of computer graphics within the UK academic community.

apE: a visualization system available from the Ohio Supercomputer Center (*see "apE 2.0" on page 194*).

API: Application Programmer's Interface (*see "Data Manipulation Languages and Macro Facilities" on page 95*).

AVS: Application Visualization System, a visualization system from Stardent Computer (*see "AVS" on page 196*).

BNF: Backus-Naur Form; a method of describing the syntax of a language.

BSI: British Standards Institute.

CAD, CAD/CAM: Computer Aided Design, Computer Aided Design / Computer Aided Manufacture.

CCTA: Central Computer and Telecommunications Agency; UK agency responsible for quality assurance in UK government data processing procurement.

CEC: Commission of the European Community.

CERN: European Centre for Nuclear Research, Geneva.

CFD: Computational Fluid Dynamics.

CGM: Computer Graphics Metafile (ISO standard 8632) [ISO(8632)87].

CGI: Computer Graphics Interface (as yet unpublished ISO standard).

CHEST: Combined Higher Education Software Trust; a UK body responsible for the provision of commercial software to UK academic institutions.

Computer Board: UK organization funding University computer centres.

Cray: Generically Cray Systems and Cray Research, firms involved in the production of supercomputers (*inter alia*).

CSG: Computational Solid Geometry.

CSSR/PLUTO: Crystal Structure Search and Retrieval, also used by the Cambridge Crystalographic Databank.

CT: Computer Tomography.

DARPA: Defense Advanced Research Projects Agency.

DataGlove: a manual input device from VPL.

DBMS: Database Management System.

DDL: Data Definition Language (*see "Data Description Languages" on page 94*).

DEC: Digital Equipment Corporation.

DML: Data Manipulation Language (*see "Data Manipulation Languages and Macro Facilities" on page 95*).

Doré: a 3D graphics system from Stardent Computer.

EC: European Community.

Eurographics, Eurographics UK: European organization for the advancement of computer graphics, and its UK chapter.

GDDM: a graphics system for IBM mainframe computers, a trademark of IBM.

GKS, GKS-3D: Graphical Kernel System (GKS); Graphical Kernel System 3D (GKS-3D); ISO standards for procedural interfaces to graphics systems.

GIF: a Graphics Interchange Format, a trademark of Compuserve Inc.

gigabit/second: a transmission rate of one thousand million bits per second.

gigabyte: slightly more than a thousand million bytes (precisely 1024 megabytes).

gigaflop: one thousand million floating point operations per second.

GINO-F: a general purpose graphics system, marketed by Bradley Associates Ltd., UK.

GIS: Geographic Information System.

GL: Graphics Library (a product from Silicon Graphics Inc).

GRASPARC: A GRAphical environment for Supporting PARallel Computing, a joint project between NAG, the University of Leeds and Quintek Ltd.

hardcopy: (in this book) the production of output on permanent, viewable media such as paper or film.

HCI: Human-Computer Interface (see chapter 5).

HDF: Hierarchical Data Format, developed at NCSA *qv.* (*see "ViSC Generic Data Formats" on page 232*).

HDTV: High Definition TeleVision; a generic term for a number of developments in Japan, Europe and the USA towards television standards using roughly double the current number of scan lines.

IBM: International Business Machines.

IEEE: Institute of Electrical and Electronic Engineers.

INGRES: an integrated relational database system designed for commercial, scientific and engineering applications; marketed by Relational Technology Inc.

ISO: International Organization for Standardization.

IT: Information Technology.

JADE: Japan Deutschland England collaborative experiment at DESY High Energy Physics Laboratory in Germany.

JPEG: Joint Photographic Experts Group; a joint ISO/CCITT project defining compression standards for still pictures (*see "Standards" on page 101*).

JNT: Joint Network Team; UK agency for the coordination of network development in the academic arena.

Khoros: A visualization system from the University of New Mexico.

Macintosh: a personal computer from Apple Computer Corporation.

Mathematica: a software product for investigating and typesetting mathematical formulae.

megabyte: slightly more than a million bytes (precisely 1,048,576 bytes).

MIDAS/FITS: data format used by astronomers.

MOVIE.BYU: a movie making system from Brigham Young University.

MPEG: Moving Picture Experts Group; a group defining a series of standards for the compression of sequences of pictures (*see "Standards" on page 101*).

MS-DOS: an operating system for personal computers.

NAG, NAG Library, NAG Graphics Library: Numerical Algorithms Group Ltd., a mathematical library from NAG, a graphical library from NAG.

NASA: National Aeronautical and Space Agency; USA body responsible for space research.

NCGA: National Computer Graphics Association.

NCSA: National Center for Supercomputer Applications, University of Illinois at Urbana-Champaign.

NERC: Natural Environment Research Council; UK body responsible for coordinating and funding research on all environmental issues.

netCDF: Network Common Data Format, developed by Unidata, part of the University Corporation for Atmospheric Research, Boulder, Colorado (*see "ViSC Generic Data Formats" on page 232*).

NSF: National Science Foundation; USA body responsible for coordinating and funding science.

NURBS: Non-Uniform Rational B-Splines; see the list of Visualization Terms earlier in this glossary.

OpenWindows: a windowing system from Sun Microsystems.

ORACLE: a database management system from Oracle Corporation.

PAW: Physics Analysis Workstation (a visualization system written at CERN, Switzerland (*see "Data Manipulation Languages and Macro Facilities" on page 95*)).

PC: Personal Computer (registered trademark of IBM).

PDB: Brookhaven Protein DataBank.

PeX: PHIGS extensions to X; extensions to the X Window System (*q.v.*) to support PHIGS (*q.v.*).

PHIGS, PHIGS PLUS: Programmer's Hierarchical Interactive Graphics Systems (ISO standard 9592); an international standard for 3D graphics [ISO/IEC(9592-1,2,3)89]; PHIGS PLUS is an extension of this [ISO/IEC(9592-4)91] - see Appendix A.

Pixel Planes: a volume rendering system from AT&T.

PostScript: a page description language, devised by and a registered trademark of Adobe Inc.

PLOT-10: a graphics system from Tektronix Inc.

PRBN: progressive recursive binary nesting, a data compression technique (*see "Compression Techniques" on page 101*).

PS/2: Personal System/2, a personal computer (registered trademark of IBM).

PV-WAVE: a visualization system from Precision Visuals Inc.

RenderMan: a rendering system from Pixar.

Research Councils: five bodies (SERC, NERC, Medical Research Council, Agriculture and Fisheries Research Council and Economic and Social Research Council) responsible for funding research in the UK.

RGB: Red Green Blue; a method of distributing colour pictures, used finally in most television-based displays.

SDSC: San Diego Supercomputer Center.

SERC: Science and Engineering Research Council, UK body coordinating and funding science and engineering research.

SGI, Silicon Graphics: Manufacturer of graphics workstations (*inter alia*).

SIGCOMM: ACM Special Interest Group in Communications.

SIGGRAPH: ACM Special Interest Group in Computer Graphics.

SparcStation: a personal workstation from Sun Microsystems.

SQL, SQL-DDL, SQL-DML: Standard Query Language; has constituent parts SQL-DDL (a DDL) and SQL-DML (a DML).

SUN, SunVision, SunView: Manufacturer of graphics workstations (*inter alia*), a visualization software product from Sun, a windowing system from Sun.

TekBase: a database management system from Leading Technology Products.

terabyte: slightly more than a million million bytes.

UNIRAS: A set of graphics libraries and program from the firm of the same name.

Unix: an operating system; a registered trademark of AT&T.

VLSI: Very Large Scale Integration, a design philosophy for complex electronic systems.

VoxelView: a volume visualization system from Vital Images Inc.

WINSOM: WINchester SOlid Modeller; a solid modelling system developed by the IBM (UK) Scientific Centre at Winchester.

X Window System, X11: a network-transparent window system developed at MIT [Nye88], [O'Reilly88].

X-server: the part of an X Window System connection seen by the user.

X toolkit: a package extending the capabilities - or improving the usability - of the basic X Window System libraries.

XDR: the External Data Representation Standard (XDR) developed by Sun Microsystems [Sun88].

XGL: a basic graphics library on Sun workstations.

X25: A networking protocol (not related to the X Window System).

Appendix C

BIBLIOGRAPHY

This bibliography consists of the full bibliographic information for all references in the book, including key references at the end of each chapter and for each section in the Applications chapter, and all references from within the body of the book.

[Abraham84]
Abraham R.H., Shaw C.D., "Dynamics. The Geometry of behavior", **(Aerial Press Inc., 1984)**.

[Adobe85]
Adobe Systems, "PostScript Language Reference Manual", **(Addison-Wesley, 1985)**.

[AGOCG90]
"PC Graphics Evaluation", *AGOCG Technical Report 3*, **(AGOCG, 1990)**.

[Ahlberg67]
Ahlberg J.H., Nilson E.N., Walsh J.L, "The Theory of Splines and their Applications", **(Academic Press, New York, 1967)**.

[Anderson89]
Anderson H.S., Berton J.A., Carswell P.G., Dyer D.S., Faust J.T., Kempf J.L., Marshall R.E., "The animation production environment: A basis for visualization and animation of scientific data", *Technical Report, Ohio Supercomputer Graphics Project,* **(Mar 1989)**.

[Andrews72]
Andrews D.F., "Plots of High Dimensional Data", *Biometrics* vol 28, *pp 125-136.*

[ANSI88]
ANSI, "PHIGS+ Functional Description Revision 3.0", **(1988)**.

[Ardent88a]
Ardent Computer Corp., "Doré Technical Overview", **(Apr 1988)**.

[Bancroft89]
Bancroft G., Plessel T., Merrit F., Watson V., "Tools for 3D visualization in computational aerodynamics at NASA Ames Research Center", *SPIE* vol 1083, **(1989)**.

[Barlow90]
"Images and Understanding: Thoughts about Images, Ideas and Understanding", ed. Barlow H., Blakemore C., Weston-Smith M., **(Cambridge University Press, 1990)**.

[Barnsley87]
Barnsley M., Sloan A., "Chaotic compression (a new twist on fractal theory speeds image transmission to video rates)", *Computer Graphics World,* **(Nov 1987)**, *pp 107-108.*

[Barnsley88]
Barnsley M., "Fractals Everywhere", **(Academic Press, New York, 1988)**.

[Beaton87]
Beaton R.J., DeHoff R.J., Weiman N., Hildebrandt P.W., "An Evaluation of Input Devices for 3D Computer Display Workstations", *True 3D Imaging and Display Technologies, SPIE* vol 761, **(1987)**, *pp 94-101.*

[Belie85]
Belie R.G., "Flow visualization in the Space Shuttle Main Engine", *Journal of Mechanical Engineering* vol 107, **(1985)**, *pp 27-33.*

[Belie86]
Belie R.G., "The Space Shuttle, Fluid Dynamics and Computer Graphics", *IEEE Computer Graphics and Applications* vol 6 (6), **(Nov 1986)**, *pp 6-7.*

[Bergeron89]
Bergeron R.G., Grinstein G.G., "A reference model for the visualization of multi-dimensional data", *Proc. Eurographics '89,* **(Elsevier Science Publishers B.V, 1989)**, *pp 393-399.*

[Bier90]
Bier E.A., "Snap-Dragging in Three Dimensions", *ACM SIGGRAPH Computer Graphics* vol 24 (2), **(Mar 1990)**, *pp 193-204.*

[Binot90]

Binot C., "Architecture and Evaluation of Graphics Superworkstations", *Eurographics '90: Tutorial Notes 6*, **(1990)**.

[Blanchard90]

Blanchard C., Burgess S., Harvill Y., Lanier J., Lasko A., Oberman M., Teitel M., "Reality Built for Two : A Virtual Reality Tool", *ACM SIGGRAPH Computer Graphics* vol 24 (2), **(Mar 1990)**, *pp 35-36*.

[Brankin89]

Brankin R., Gladwell I., "Shape-preserving local interpolation for plotting solution of ODEs", *IMA Journal of Numerical Analysis* vol 9, **(Academic Press, 1989)**, *pp 555-566*.

[Brodlie91]

Brodlie K.W., Butt S, "Preserving convexity using piecewise cubic interpolation", *Computers and Graphics* vol 14 (4), **(1991)**.

[Brown80]

Brown W.S., "A Simple but Realistic Model for Floating-Point Computation", *TOMS* vol 6, **(Dec 1980)**, *pp 510-523*.

[Brown88]

Brown I.D., "Standard Crystallographic File Structure", *Acta Crystallographica* vol 44, **(1988)**, *pp 232*.

[Bruce85]

Bruce V., Green P., "Visual Perception: Physiology, Psychology and Ecology", **(Lawrence Erlbaum Associates, 1985)**.

[Brun89]

Brun R., Couci O., Vandoni C.E., Zanari P., "PAW, a general purpose portable software tool for data analysis and presentation", *Proc. Conference on Computing in High-Energy Physics*, **(Oxford, 1989)**.

[Buning88]

Buning P.G., "Sources of Error in the Graphical Analysis of CFD Results", *Journal of Scientific Computing* vol 3 (2), **(1988)**, *pp 149-164*.

[Card83]

Card S.K., Moran T.P., Newell A., "The Psychology of Human-Computer Interaction", **(Lawrence Erlbaum Associates, 1983)**.

[Carlson85]

Carlson R.E., Fritsch F.N., "Monotone piecewise bicubic interpolation", *SIAM Journal of Numerical Analysis* vol 22 (2), **(1985)**, *pp 386-400.*

[Carpenter91]

Carpenter L.A., "The Visualisation of Numerical Computation", *Eurographics Workshop on Scientific Visualization*, **((to be published 1991), 1990)**.

[CG90]

"Special Issue on San Diego Workshop on Volume Visualisation", *Computer Graphics* vol 24 (5), **(Nov 1990)**.

[Chang89]

Chang R.E., "Beyond the Third Dimension", *IRIS Universe* vol Fall 89, **(1989)**, *pp 26-33.*

[Charalamides90]

Charalamides S., "New Wave technical graphics is welcome", *DEC USER*, **(Aug 1990)**, *pp 49-50.*

[Chen85]

Chen L., Herman G.T., Reynolds R.A., Udupa J.K., "Surface Shading in the Cuberille Environment", *IEEE Computer Graphics and Applications* vol 5 (6), **(Dec 1985)**, *pp 33-43.*

[Chernoff73]

Chernoff H., "The use of faces to represent points in k-dimensional space graphically.", *Journal of the American Statistical Society* vol 68 (342), **(1973)**, *pp 361-368.*

[Clifford88]

Clifford W.H. Jr., McConnell J.I., Saltz J.S., "The Development of PEX, a 3D Extension to X11", *Proc. Eurographics '88*, ed. Duce D.A., Jancene P., **(North-Holland)**.

[Cline88]

Cline H.E., Lorensen W.E., Ludke S., Crawford C.R., Teeter B.L., "Two Algorithms for the Three-Dimensional Reconstruction of Tomograms", *Medical Physics* vol 15 (3), **(May / June 1988)**, *pp 320-327.*

[Collins90]

Collins B.M., Phippen R.W., Quarendon P., Watson D., Whitfield G.A., Williams D.W., Wyatt M.J., "WinVis90 or a Mathematical Visualiser", *IBM UKSC Report No 227,* **(Apr 1990)**.

[CompPhys90]

"Special Issue on Chaos", *Computer in Physics* vol 4 (5), **(1990)**.

[Coonen84]

Coonen J.T., "Contributions to a Proposed Standard for Binary Floating-Point Arithmetic", **(PhD Thesis, University of California, Berkeley, 1984)**.

[Cox84]

Cox M.D., "A Primitive Equation: 3-Dimensional Model of the Ocean", *Geophysical Fluid Dynamics Laboratory Ocean Group Technical Report 1,* **(Princeton, 1984)**.

[Crawford90]

Crawford S.L., Fall T.C., "Projection Pursuit Techniques for the Visualization of High Dimensional Datasets", *Visualisation in Scientific Computing,* ed. Nielson G.M., Shriver B., Rosenblum L.J., **(IEEE Computer Society Press, 1990)**, *pp 94-108.*

[Crennell89]

Crennell K.M., Carter M.L., Golton E., Maybury R., Bartlett A., Oldfield R., Hammarling S., "The Design and Implementation of a portable Image Processing Algorithms Library (IPAL) in Fortran and C", *Proc. Third IEE International Conference on Image Processing and its Application,* ISBN 0-85296-382-3, **(Warwick, July 1989)**, *pp 516-520.*

[Date81]

Date C.J., "An Introduction to Database Systems", **(Addison-Wesley, 1981)**.

[DeFanti89]
DeFanti T.A., Brown M.D., McCormick B.H., "Visualization - Expanding Scientific and Engineering Research Opportunities", *IEEE Computer* vol 23 (8), **(Aug 1989)**, *pp 12-25.*

[Drebin88]
Drebin R.A., Carpenter L., Hanrahan P., "Volume Rendering", *ACM SIGGRAPH Computer Graphics* vol 22 (4), **(Aug 1988)**, *pp 65-74.*

[Duff81]
"Languages and architectures for image processing", ed. Duff M.J.B., Levialdi S., **(Academic Press, 1981)**.

[Durrett87]
"Colour and the Computer", ed. Durrett J.H., **(Academic Press, 1987)**.

[Dyer90]
Dyer D.S., "A Dataflow Toolkit for Visualization", *IEEE Computer Graphics and Applications* vol 10 (4), **(July 1990)**, *pp 60-69.*

[Eason88]
Eason K., "Information Technology and Organisational Change", **(Taylor and Francis, 1988)**.

[Eaton87]
Eaton B.I., "Analysis of Laminar Vortex Shedding behind a Cylinder by Computer Aided Flow Visualization", *Journal of Fluid Mechanics* vol 180, **(1987)**, *pp 117-145.*

[Edwards89]
Edwards D.E., "3 dimensional Visualisation of Fluid dynamics", *AIAA Paper 89-0136*, **(1989)**.

[Elvins90]
Elvins T.T., "A Visualization Computing Environment for a Widely Dispersed Scientific Community", *State of the Art in Data Visualization: SIGGRAPH Course Notes*, **(Aug 1990)**.

[Everitt78]
Everitt B., "Graphical techniques for multivariate data.", **(Heinemann, 1978)**.

[Fairchild88]

Fairchild K.M., Poltrock S.E., Furnas G.W., "Semnet: 3D graphics representations of large knowledge bases", *Cognitive Science and Its Applications for Human Computer Interaction*, ed. Guindon R., **(Lawrence Erlbaum Associates, 1988)**.

[Foley90a]

Foley J.D., van Dam A., Feiner S.K., Hughes J.F., "Computer Graphics, Principles and Practice", **(Addison-Wesley, 1990)**.

[Foley90b]

Foley T.A., Lane D.A., Nielson G.M., Ramaraj R., "Visualizing Functions Over a Sphere", *IEEE Computer Graphics and Applications* vol 10 (1), **(Jan 1990)**, *pp 32-40.*

[Frenkel88]

Frenkel K. A., "The Art and Science of Visualizing Data", *Communications of the ACM* vol 31 (2), **(1988)**, *pp 110-121.*

[Fritsch80]

Fritsch F.N., Carlson R.E., "Monotone piecewise cubic interpolation", *SIAM Journal of Numerical Analysis* vol 17, **(1980)**, *pp 238-246.*

[Frolich89]

Frolich R., "The Shelf Life of Antarctic Ice", *New Scientist*, **(Nov 4, 1989)**, *pp 62-65.*

[Fuchs77]

Fuchs H., Kedem Z.M., Uselton S.P., "Optimal Surface Reconstruction from Planar Contours", *Communications of the ACM* vol 20 (10), **(1977)**, *pp 693-702.*

[Fuchs89a]

Fuchs H., Levoy M., Pizer S.M., "Interactive Visualization of 3D Medical Data", *IEEE Computer*, **(Aug 1989)**, *pp 46-51.*

[Fuchs89b]

Fuchs H., Levoy M., Pizer S.M., Rosenman J.G., "Interactive Visualization and Manipulation of 3-D Medical Image Data", *Proc. NCGA '89 Conference Volume 1*, **(1989)**.

[Fuller80]

Fuller A.J., dos Santos M.L.X., "Computer Generated Display of 3D Vector Fields", *Computer Aided Design* vol 12 (2), **(1980)**, *pp 61-66.*

[Fung65]

Fung Y.C., "Foundations of Solid Mechanics", **(Prentice Hall, 1965)**.

[Gallop90]

Gallop J.R., Haswell J., Maybury R., Popovic R., Thomas R.E., "Assessment of Superworkstations for EASE", *EASE Technical Report*, **(Rutherford Appleton Laboratory, Nov 1990)**.

[Getzels80]

Getzels J.W., "The Psychology of Creativity", *Carnegie Symposium on Creativity (Inaugural Meeting of the Library of Congress Council of Scholars)*, **(Nov 1980)**.

[GIF87]

"Graphics Interchange Format - A standard defining a mechanism for the storage and transmission of raster-based graphics information", **(CompuServe Incorporated, 1987)**.

[Goosens89]

"PAW - Physics Analysis Workstation. The complete reference, version 1.07", *CERN Program Library Entry Q 121*, ed. Goosens M., **(Oct 1989)**.

[Gosling89]

Gosling J., Rosenthal D.S.H., Arden M.J., "The NeWS Book", **(Springer-Verlag, 1989)**.

[Gregory70]

Gregory R.L., "The Intelligent Eye", **(Weidenfeld and Nicholson, 1970)**.

[Gregory77]

Gregory R.L., "Eye and Brain: The Psychology of Seeing", **(Weidenfeld and Nicholson, 1977)**.

[Grimsrud89]

Grimsrud A., Lorig G., "Implementing a Distributed Process between Workstation and Supercomputer", *Proc. First International Conference on Applications of Supercomputers in Engineering: Algorithms, Computer Systems and User Experience,* ed. Brebbia C.A., Peters A., **(1989)**.

[Haber90]

Haber R.B., McNabb D.A., "Visualization Idioms : A Conceptual Model for Scientific Visualization Systems", *Visualization in Scientific Computing,* ed. Nielson G.M., Shriver B., Rosenblum L.J., **(1990)**, *pp 74-93.*

[Haeberli88]

Haeberli P.E., "ConMan : A Visual Programming Language for Interactive Graphics", *ACM SIGGRAPH Computer Graphics* vol 22 (4), **(Aug 1988)**, *pp 103-112.*

[Hamming62]

Hamming R.W., "Numerical Methods for Scientists and Engineers", **(McGraw-Hill, New York, 1962)**.

[Haswell90]

Haswell J., "The suitability of current visualization systems for real applications", **(MSc thesis, Middlesex Polytechnic, 1990)**.

[Hayes74]

Hayes J.G., Halliday J., "The Least Squares Fitting of Cubic Spline Surfaces to General Data Sets", *Journal of the Institute of Mathematics and its Applications* vol 14, *pp 89-103.*

[Hayes79]

Hayes J., "Cognitive Psychology and Interaction", *Methodology of Interaction,* ed. Guedj R.A., ten Hagen P.J.W., Hopgood F.R.A., Tucker H.A., Duce D.A, ISBN 0-444-85479-7, **(Elsevier North-Holland, 1979)**.

[Helman89]

Helman J., Hesselink L., "Automated Analysis of Fluid Flow Topology", *Three-Dimensional Visualisation and Display Technologies, SPIE* vol 1083, **(1989)**, *pp 144-152.*

[Helman90]

Helman J., Hesselink L., "Representation and Display of Vector Field Topology in Fluid Flow Data Sets", *Visualization in Scientific Computing,* ed. Nielson G.M., Shriver B., Rosenblum L.J., **(IEEE Computer Society Press, 1990)**, *pp 61-73.*

[Herman79]

Herman G.T., Liu H.K., "Three dimensional display of organs from computed tomograms", *Computer Graphics and Image Processing* vol 9, *pp 1-12.*

[Herman83]

Herman G.T., Udupa J.K., "Display of 3D Digital Images : Computational Foundations and Medical Applications", *IEEE Computer Graphics and Applications* vol 3 (5), **(Aug 1983)**, *pp 39-46.*

[HMSO80]

HMSO, "People in Britain - a Census Atlas", **(Census Research Unit/ Office of Population Censuses and Surveys/ General Registrars Office (Scotland), 1980)**.

[Hibbard89]

Hibbard W., Santek D., "Interactivity is the Key", *Proc. Chapel Hill Workshop on Volume Visualisation,* **(May 1989)**, *pp 39-43.*

[Hopgood91]

Hopgood F.R.A, "Using Colour in Computer Graphics", *AGOCG Technical Report 4,* **(AGOCG, Jan 1991)**.

[Hopkins90]

Hopkins T.R., "NAG Spline Fitting Routines on a Graphics Workstation - the Story so far", *NAG Newsletter '90* vol 2, **(1990)**, *pp 11-17.*

[Hu89]

Hu X., Tan K.K., Levin D.N., Galhotra S.G., Pelizzari C.A., Chen G.T.Y., Beck R.N., Chen C-T., Cooper M.D., "Volumetric Rendering of Multi-Modality, Multivariable Medical Imaging Data", *Proc. Chapel Hill Workshop on Volume Visualisation,* **(May 1989)**, *pp 45-49.*

[Huang79]
Huang T.S., "Picture Processing and Digital Filtering - Introduction", *Topics in Applied Physics (Second Edition)* vol 6, **(Springer-Verlag, 1979)**.

[Hubbold90]
Hubbold R.J., "Interactive Scientific Visualisation: a Position Paper", *Eurographics Workshop on Visualisation in Scientific Computing*, **(1990)**.

[Hudson87]
Hudson G.P., "Photovideotex image compression algorithms - towards international standardisation", *ESPRIT'87 Achievements and Impact Part 2*, **(North-Holland Elsevier, 1987)**, *pp 1137-1148.*

[IEEE85]
"Standard for Binary Floating Point Arithmetic", *ANSI/IEEE 754-1985.*

[IOC84]
"GF3: IOC General magnetic tape format for international exchange of oceanographic data", *IOC Manuals and Guides No.9, Annex 1, part 3 introductory Guide*, **(UNESCO, 1984)**.

[ISO(7942)85]
ISO, "Information processing systems - Computer Graphics - Graphical Kernel System (GKS) functional description", **(1985)**.

[ISO(8632)87]
ISO, "Information processing systems - Computer Graphics - Metafile for transfer and storage of picture description information", **(1987)**.

[ISO(9075)87]
ISO, "Information processing systems - Database Language SQL", **(1987)**.

[ISO(8651-1)88]
ISO, "Information processing systems - Computer Graphics - Graphical Kernel System (GKS) language bindings - Part 1 : Fortran", **(1988)**.

[ISO(8651-2)88]
ISO, "Information processing systems - Computer Graphics - Graphical Kernel System (GKS) language bindings - Part 2 : Pascal", **(1988)**.

[ISO(8651-3)88]
ISO, "Information processing systems - Computer Graphics - Graphical Kernel System (GKS) language bindings - Part 3 : Ada", **(1988)**.

[ISO(8805)88]
ISO, "Information processing systems - Computer Graphics - Graphical Kernel System for Three Dimensions (GKS-3D) functional description", **(1988)**.

[ISO(8806-1)88]
ISO DIS, "Information processing systems - Computer Graphics - Graphical Kernel System for Three Dimensions (GKS-3D) language bindings - Part 1 : Fortran", **(1988)**.

[ISO(8806-3)88]
ISO DIS, "Information processing systems - Computer Graphics - Graphical Kernel System for Three Dimensions (GKS-3D) language bindings - Part 3 : Ada", **(1988)**.

[ISO/IEC(9592-1,2,3)89]
ISO/IEC, "Information processing systems - Computer Graphics - Programmer's Hierarchical Interactive Graphics System - Parts 1, 2 and 3", **(1989)**.

[ISO(8651-1)90]
ISO - DIS, "Information processing systems - Computer Graphics - Graphical Kernel System (GKS) language bindings - Part 4 : C", **(1990)**.

[ISO(8806-4)90]
ISO DIS, "Information processing systems - Computer Graphics - Graphical Kernel System for Three Dimensions (GKS-3D) language bindings - Part 4 : C", **(1990)**.

[ISO/IEC(9593-1)90]
ISO/IEC, "Information processing systems - Computer Graphics - Programmer's Hierarchical Interactive Graphics System - Language bindings : Part 1 : Fortran", **(1990)**.

[ISO/IEC(9593-3)90]
ISO/IEC, "Information processing systems - Computer Graphics - Programmer's Hierarchical Interactive Graphics System - Language bindings : Part 3 : Ada", **(1990)**.

[ISO/IEC(9593-4)90]
ISO/IEC DIS, "Information processing systems - Computer Graphics - Programmer's Hierarchical Interactive Graphics System - Language bindings : Part 4 : C", **(1990)**.

[ISO/IEC(9592Am1)91]
ISO/IEC, "Information processing systems - Computer Graphics - Programmer's Hierarchical Interactive Graphics System - Amendment 1 to Parts 1, 2 and 3", **(1991)**.

[ISO/IEC(9592-4)91]
ISO/IEC DIS, "Information processing systems - Computer Graphics - Programmer's Hierarchical Interactive Graphics System - Part 4 : Plus Lumière und Surfaces, PHIGS PLUS", **(1991)**.

[ISO(10967)91]
ISO - CD, "Information technology - Programming languages - Language compatible arithmetic", **(1991)**.

[ISO91]
"Proposal for a New Work Item: Image Processing and Interchange", **(ISO/IEC JTC 1/SC 24, Jan 1991)**.

[Jain89]
Jain A.K., "Fundamentals of Digital Image Processing", **(Prentice Hall International, 1989)**.

[Jern89]
Jern M., "Visualisation of Scientific Data"; *Computer Graphics 89*, ISBN 0-86363-190-3, **(Blenheim On-Line, 1989)**, *pp 79-103*.

[JMG85]
"Standard Crystallographic File Structures-84", *Journal Molecular Graphics* vol 3 (2), **(June 85)**, *pp 40*.

[Kaufman90]
Kaufman A., "3D Volume Visualization", *Eurographics '90: Tutorial Notes 12*, **(1990)**.

[Kenada89]
Kenada K., Kato F., Nakamae E., Nishita T., "Three dimensional Terrain Modelling and Display for Environmental assessment", *Computer Graphics* vol 23 (3), **(1989)**, *pp 207-214.*

[Keppel75]
Keppel E., "Approximating Complex Surfaces by Triangulation of Contour Lines", *IBM Journal of Research and Development* vol 19 (1), **(Jan 1975)**, *pp 1-96.*

[Kroos85]
Kroos K.A., "Computer Graphics Techniques for Three-Dimensional Flow Visualization", *Frontiers in Computer Graphics,* ed. Kunii T.L., **(Springer-Verlag, New York, 1985)**.

[Lancaster86]
Lancaster P., Salkauskas K, "Curve and Surface Fitting - An Introduction", **(Academic Press, 1986)**.

[Lang90]
Lang U., Aichele H., Pöhlmann H., Rühle R., "Scientific Visualization in a Supercomputer Network", *Eurographics Workshop on Scientific Computing 1990,* **((to be published), 1991)**.

[Lansdown89]
"Computers in Art, Design and Animation", ed. Lansdown R.J., Earnshaw R.A., **(Springer-Verlag, 1989)**.

[LeGall91]
Le Gall D., "MPEG: A Video Compression Standard for Multimedia Applications", *Communications of the ACM,* **(Apr 1991)**, *pp 46-58.*

[Levoy88]
Levoy M., "Display of Surfaces from Volume Data", *IEEE Computer Graphics and Applications* vol 8 (3), **(May 1988)**, *pp 29-37.*

[Levoy90a]
Levoy M., Whittaker R., "Gaze-directed Volume Rendering", *Special Issue on the 1990 Symposium on Interactive 3D Graphics, ACM SIGGRAPH Computer Graphics* vol 24 (2), **(Mar 1990)**, *pp 217-223.*

[Levoy90b]
Levoy M., "A Taxonomy of Volume Visualization Algorithms", *Introduction to Volume Visualization: SIGGRAPH Course Notes,* **(Aug 1990)**, *pp 6-8.*

[Little87]
Little C.T., "Graphics at the UK Met Office", *Proc. BCS Conference on The future of graphics software,* ed. Earnshaw R.A., **(1987)**.

[Lorensen87]
Lorensen W.E., Cline H.E., "Marching Cubes : A High Resolution Surface Construction Algorithm", *ACM SIGGRAPH Computer Graphics* vol 21 (4), **(July 1987)**, *pp 163-169.*

[MacDonald90]
MacDonald L.W., "Using Colour Effectively in Displays for Computer-Human Interface", *Displays,* **(July 1990)**, *pp 129-141.*

[Maguire90]
Maguire D.J., Worboys M.F., Hearnshaw H.M., "An Introduction to Object-Oriented Geographical Information Systems", *Mapping Awareness* vol 4 (2), **(1990)**, *pp 36-39.*

[Maybury90]
Maybury R., Bartlett A.D.H., "The NAG/SERC Image Processing Algorithms Library (IPAL)", *RAL Report RAL-90-051,* **(Rutherford Appleton Laboratory)**.

[McCormick87]
McCormick B., DeFanti T.A., Brown M.D., "Visualization in Scientific Computing", *ACM SIGGRAPH Computer Graphics* vol 21 (6), **(Nov 1987)**.

[Meagher84]
Meagher D.J., "Interactive Solids Processing for Medical Analysis and Planning", *Proc. 5th NCGA Conference 1984,* **(NCGA)**, *pp 96-106.*

[Mitchie85]
Mitchie A., Aggarwal J.K., "Image Segmentation by Conventional and Information Integrating Techniques: - A Synopsis", *Image and Vision Computing* vol 3 (2), **(May 1985)**, *pp 50-62.*

[MO87]
Meteorological Office, "The Storm of 15/16 Oct 1987", *Met Office Report*, ISBN 0861-80-2322.

[Monk84]
"Fundamentals of Human-Computer Interaction", ed. Monk A., **(Academic Press, 1984)**.

[Morffew85]
Morffew A.J., "Protein Modelling using Computer Graphics", *Advances in Biotechnological Processes 5,* **(1985)**, *pp 31-58.*

[Murch86]
Murch G.M., "Human Factors of Colour Displays", *Advances in Computer Graphics,* ed. Hopgood F.R.A, Hubbold R.J., Duce D.A., **(Springer-Verlag, 1986)**.

[Murch89]
Murch G.M., "Colour in Computer Graphics: Manipulating and Matching Colour", *Advances in Computer Graphics,* ed. Purgathofer W., Schoenhut J., **(Springer-Verlag, 1989)**.

[Myers90]
Myers B.A., "Taxonomies of visual programming and program visualization", *Journal of Visual Languages and Computing* vol 1 (1), **(1990)**, *pp 97-123.*

[NAG90]
NAG, "NAG Library Manual", **(NAG Ltd, Oxford, 1990)**.

[NCSA89]
"NCSA, HDF Calling Interfaces and Utilities", *NCSA HDF Version 3.1,* **(National Center for Supercomputing Applications at the University of Illinois Urbana-Champaign, Mar 1989)**.

[Nielson90]
"Visualization in Scientific Computing", ed. Nielson G.M., Shriver B., Rosenblum L.J., ISBN 0-8186-8979-X, **(IEEE Computer Society Press, 1990)**.

[Nye88]
Nye A., "Xlib Programming Manual" vol 1, **(O'Reilly & Associates Inc., 1988)**.

[Nyquist28]
Nyquist H., "Certain topics in telegraph transmission theory", *AIEE Transactions* vol 47, **(1928)**, *pp 617-644*.

[OED69]
"Oxford English Dictionary", **(Oxford University Press, 1969)**.

[O'Reilly88]
O'Reilly, "Xlib Reference Manual" vol 2, **(O'Reilly & Associates Inc., 1988)**.

[Pearson91]
"Image Processing", ed. Pearson D.E., **(McGraw-Hill, 1991)**.

[Pickover88]
Pickover C., "The Use of Image Processing Techniques in Rendering Maps with Deterministic Chaos", *The Visual Computer* vol 4, **(1988)**, *pp 271-276*.

[Pickover90]
Pickover C., "Computer Pattern, Chaos and Beauty. Graphics from an Unseen World", **(St Martins Press, New York, 1990)**.

[Pixar88]
Pixar, "The RenderMan Interface Version 3.0", **(Pixar, 1988)**.

[Poskanzer89]
Poskanzer J., "Extended Portable Bitmap Toolkit", **(available on X11.4 release tape from MIT, 1989)**.

[Powell77]
Powell M.J.D., Sabin M., "Piecewise quadratic approximation on triangles", *ACM Transactions on Mathematical Software* vol 3, **(1977)**, *pp 316-325*.

[Pratt78]
Pratt W.K., "Digital Image Processing", **(John Wiley, 1978)**.

[Preusser86]
Preusser A., "Computing area-filled contours for surfaces defined by piecewise polynomials.", *Computer Aided Geometric Design* vol 3, **(1986)**, *pp 267-279*.

[Preusser89]
Preusser A., "ALGORITHM 671 - FARB-E-2D: Fill Area with Bicubics on Rectangles - A Contour Plot Program", *ACM Transactions on Mathematical Software* vol 15 (1), **(1989)**, *pp 79-89.*

[Purvis86]
Purvis G.D., Culberson C., "On the Graphical Display of Molecular Electrostatic Force-Fields and Gradients of the Electron Density", *L Molecular Graphics*, **(June 1986)**, *pp 89-92.*

[Quarendon84]
Quarendon P., "WINSOM User's Guide", *IBM UKSC report 124*, **(1984)**.

[Rasure91]
Rasure J., Argiro D., Sauer T., Williams C., "A Visual Language and Software Development Environment for Image Processing", *International Journal of Imaging Systems and Technology*, **(1991)**.

[Reilly89]
Reilly P., "Data Visualization in Archaeology", *IBM System Journal* vol 28 (4), **(1989)**.

[Reilly90]
"Communication in Archaeology : a global view of the impact of information technology Volume One : Data Visualization", *Proc. Second World Archaeological Conference*, ed. Reilly P., Rahtz S., **(July 1990)**.

[Renka84]
Renka R.J., Cline A.K., "A triangle based C1 interpolation method", *Rocky Mountain Journal of Maths* vol 14, **(1984)**, *pp 223-237.*

[Renka88]
Renka R.J., "Multivariate interpolation of large sets of data", *ACM Transactions on Mathematical Software* vol 14, **(1988)**, *pp 139-148.*

[Rew90a]
Rew R.K., Davis G., "NetCDF : An Interface for Scientific Data Access", *IEEE Computer Graphics and Applications* vol 10 (4), **(July 1990)**, *pp 76-82.*

[Rew90b]

Rew R.K., "NetCDF User's Guide", *NCAR Technical Note NCAR/TN-334+1A*, **(Unidata Program Center, Boulder, Colorado, June 1990)**.

[Richter90]

Richter R., Vos J.B., Bottaro A., Gavrilakis S, "Visualization of flow simulations", *Scientific Visualization and Graphics Simulation*, ed. Thalmann D., **(John Wiley, 1990)**, *pp 161-171*.

[Robertson86]

Robertson P.K., O'Callaghan J.F., "The Generation of Colour Sequences for Univariate and Bivariate Mapping", *IEEE Computer Graphics and Applications* vol 6 (1), **(Feb 1986)**, *pp 24-32*.

[Robertson88]

Robertson P., "Visualizing Colour Spaces : A User Interface for the Effective Use of Perceptual Colour in Data Displays", *IEEE Computer Graphics and Applications* vol 8 (5), **(September 1988)**, *pp 50-64*.

[Robinson90]

Robinson P., "Stereo 3D", *Computer Graphics World*, **(June 1990)**.

[Rosenfeld82]

Rosenfeld A., Kak A.C., "Digital picture processing (Second edition)", **(Academic Press, 1982)**.

[Rosenfeld84]

Rosenfeld A., "Multiresolution Image Processing and Analysis", **(Springer-Verlag, 1984)**.

[Sabin86]

Sabin M.A., "A Survey of Contouring Methods", *Computer Graphics Forum* vol 5, *pp 325-339*.

[Samet88a]

Samet H., Webber R.E., "Hierarchical data structures and algorithms for computer graphics, Part i: Fundamentals", *IEEE Computer Graphics and Applications*, **(May 1988)**, *pp 48-68*.

[Samet88b]

Samet H., Webber R.E., "Hierarchical data structures and algorithms for computer graphics, Part ii: Advanced Applications", *IEEE Computer Graphics and Applications,* **(July 1988)**, *pp 59-75.*

[Sauer90]

Sauer T., Rasure J., Gage C., "Near Ground Level Sensing for Spatial Analysis of Vegetation", *IAPR TC7 Workshop on Multi-source Data Integration in Remote Sensing,* **(June 1990)**.

[Seum89]

Seum C.S., Wilcox R.W., "Aspects of AMIGAS II Design and Implementation", *Proc. Second ECMWF Workshop on Meteorological Operational Systems,* **(Dec 1989)**, *pp 121-125.*

[Sewell88]

Sewell G.E., "Plotting Contour Surfaces of a Function of Three Variables.", *ACM Transactions on Mathematical Software* vol 14 (1), **(Mar 1988)**.

[Shannon49]

Shannon C.E., "Communications in the presence of noise", *Proc IRE* vol 37, **(Jan 1949)**, *pp 10-21.*

[Shirley89]

Shirley P., Neeman H., "Volume Visualisation at the Centre for Supercomputing Research and Development", *Proc. Chapel Hill Workshop on Volume Visualisation,* **(May 1989)**, *pp 17-21.*

[Shneiderman87]

Shneiderman B., "Designing the User Interface", **(Addison-Wesley, 1987)**.

[Spatial89a]

Spatial Systems Inc., "Spaceball Technical Reference V2.0", **(1989)**.

[Spatial89b]

Spatial Systems Inc., "Spaceball Application Developers Reference V4.0", **(1989)**.

[Stackpole89]
Stackpole J.D., "GRIB and BUFR: The only codes you will ever need", *Proc. Second ECMWF Workshop on Meteorological Operational Systems,* **(1989)**, *pp 44-67.*

[Sun88]
"External Data Representation Standard: Protocol Specification", *Network Programming Manual,* **(Sun Microsystems Inc., 1988)**, *pp 127-142.*

[Sun90]
"SunVision, Sun's Visualization Software Platform", *Technical White Paper,* **(Sun Microsystems Inc.,, Mar 1990)**.

[Takamura83]
Takamura H. et al, "The design and Implementation of SPIDER - a transportable image processing software package", *Computer Vision, Graphics and Image Processing* vol 23, **(1989)**, *pp 273-294.*

[Tam88]
Tam Y-W., Davis W.A., "Display of 3D Medical Images", *Proc. Graphics Interface 1988, pp 78-86.*

[Thalmann90]
"Scientific Visualization and Graphics Simulation", ed. Thalmann D., ISBN 0-471-92742-2, **(John Wiley, 1990)**.

[Thalmann91]
"New Trends in Animation and Visualization", ed. Thalmann D., Magnenat-Thalmann N., **(John Wiley, 1991)**.

[Tiede90]
Tiede U., Höhne K-H., Bomans M., Pommert A., Riemer M., Wiebecke G., "Investigation of Medical 3D-Rendering Algorithms", *IEEE Computer Graphics and Applications* vol 10 (2), **(Mar 1990)**, *pp 41-53.*

[TOG86]
"Special Issue on User Interface Design", *ACM Transactions on Graphics* vol 5(2), 5(3), 5(4), **(1986)**.

[Tufte83]
Tufte E.R., "The Visual Display of Quantitative Information", **(Graphics Press, 1983)**.

[Upson89a]

Upson C., Faulhaber T., Kamins D., Laidlaw D., Schlegel D., Vroom J., Gurwitz R., van Dam A., "The Application Visualization System : A Computational Environment for Scientific Visualization", *IEEE Computer Graphics and Applications* vol 9 (4), **(July 1989)**, *pp 30-42.*

[Upson89b]

Upson C., "Scientific Visualization Environments for the Computational Sciences", *Proc. Compcon 1989, pp 322-327.*

[Upson91]

Upson C., "Volumetric Visualization Techniques", *State of the Art in Computer Graphics - Visualization and Modelling,* ed. Rogers D.F., Earnshaw R.A., **(Springer-Verlag, 1991)**.

[Upstill90]

Upstill S., "The RenderMan Companion", **(Addison-Wesley, 1990)**.

[vanderLans89]

van der Lans R.F., "The SQL standard - a complete reference", **(Prentice Hall International, 1989)**.

[vandeWettering90]

van de Wettering M., "The Application Visualization System - AVS 2.0", *Pixel,* **(July / Aug 1990)**.

[Vandoni89]

Vandoni C.E., "Development of a Large Graphics-based Application Package", *Computers and Graphics* vol 13 (2), **(1989)**, *pp 243-252.*

[Visvalingam81]

Visvalingam M, "The signed chi-score measure for the classification and mapping of polychotomous data", *Cartographic Journal* vol 18 (1), **(1981)**, *pp 32-43.*

[Visvalingam90]

"The Douglas-Peucker algorithm for line simplification : re-evaluation through visualization", ed. Visvalingam M., J.D.Whyatt, *Computer Graphics Forum* vol 9 (3), **(1990)**, *pp 213-228.*

[VoxelView89]
"VoxelView (Version 1.1) User's Guide", **(Vital Images Inc, 1989)**.

[Wallace91]
Wallace G.K., "The JPEG Still Picture Compression Standard", *Communications of the ACM,* **(Apr 1991)**, *pp 30-44.*

[Wallas26]
Wallas G., "The Art of Thought", **(Jonathan Cape, London, 1926)**.

[Wang90]
Wang S-L.C., Staudhammer J., "Visibility Determination on Projected Grid Surfaces", *IEEE Computer Graphics and Applications,* **(July 1990)**, *pp 36-43.*

[Ware88]
Ware C., Jessome D.R., "Using the BAT : A Six-Dimensional Mouse for Object Placement", *IEEE Computer Graphics and Applications* vol 8 (6), **(Nov 1988)**, *pp 65-70.*

[Warnock69]
Warnock J., "A Hidden-Surface Algorithm for Computer Generated Half-Tone Pictures", *Technical Report TR 4-15, NTIS AD-753 671,* **(Computer Science Department, University of Utah, June 1969)**.

[Watkins87]
Watkins H.K., "Graphics in Reservoir Simulation", *Computer Graphics Forum* vol 6, **(1987)**, *pp 111-118.*

[Watson90]
Watson D., "The State of the Art of Visualisation", *Proc. SuperComputing Europe Fall Meeting, Aachen,* **(Sept 90)**.

[Webber90]
Webber R., "Ray Tracing Voxel Data via Biquadratic Local Surface Interpolation", *The Visual Computer* vol 6 (1), **(Feb 1990)**, *pp 8-15.*

[Webster70]
"Webster's New World Dictionary", **(Collins, 1970)**.

[Wells81]
Wells D.C., Greisen E.W., Harten R.H., "FITS : A Flexible Image Transport System", *Astronomy and Astrophysics Supplement Series* vol 44, **(1981)**, *pp 363-370.*

[Williams88]
Williams R.D., Garcia F., "A Real-Time Autostereoscopic Multiplanar 3D Display System", *Society for Information Display Digest*, **(May 1988)**, *pp 91-94.*

[Williams90]
Williams C., Rasure J., "A Visual Language for Image Processing", *Workshop on Visual Languages, Skokie, Illinois*, **(IEEE Computer Society Press, Oct 1990)**.

[WMO89]
"WMO Manual of Codes, No.306, Geneva, 1989 and Supplement No 1, 1989", **(ECMWF, Mar 1988)**.

[Wright89]
Wright J., "Altered States: A software developer's vision of the future of virtual reality", *Computer Graphics World* vol 12 (12), **(Dec 1989)**.

[XDS89]
"NCSA X DataSlice", **(NCSA, 1989)**.

[XImage89]
"NCSA X Image for the X window System, Version 1.02", **(NCSA, Nov 1989)**.

[Yannakoudakis88]
Yannakoudakis E.J., "The Architectural Logic of Database Systems", **(Springer-Verlag, 1988)**.

[Ziv77]
Ziv J., Lempel A., "A universal algorithm for sequential data compression", *IEEE Transaction of Information Theory* vol IT-23 (3), **(1977)**, *pp 337-343.*

[Ziv78]
Ziv J., Lempel A., "Compression of Individual Sequences via Variable-Rate Coding", *IEEE Transaction of Information Theory* vol IT-24 (5), **(1978)**, *pp 530-536.*

Index

Printing: Mercedesdruck, Berlin
Binding: Buchbinderei Lüderitz & Bauer, Berlin